PREFACE

In March 1987, the International Commission on Radiological Protection (ICRP) established a Task Group of Committee 2 on Age-dependent Dosimetry. The Task Group was given responsibility for developing age-specific dosimetric and biokinetic models for selected elements and radionuclides. The Task Group, now renamed the Task Group on Internal Dosimetry, together with the Task Group on Dose Calculations, has prepared four publications giving dose coefficients for members of the public.

In the first report, *ICRP Publication 56* (ICRP, 1989), age-specific biokinetic models and ingestion dose coefficients (committed dose equivalents to tissues and organs and committed effective dose equivalents) were given for intakes of selected radioisotopes of hydrogen, carbon, strontium, zirconium, niobium, ruthenium, iodine, caesium, cerium, plutonium, americium, and neptunium. The calculation of committed effective dose equivalent was based on the tissue weighting factors, w_T, given in *ICRP Publication 26* (ICRP, 1977). The report also gave preliminary age-dependent inhalation dose coefficients for radioisotopes of these elements using the lung model described in *ICRP Publication 30* (ICRP, 1979).

In *ICRP Publication 67* (ICRP, 1993), age-specific biokinetic models were given for sulphur, cobalt, nickel, zinc, molybdenum, technetium, silver, tellurium, and polonium. A generic model for the biokinetics of the alkaline earth elements strontium, barium, and radium was introduced. This model was also applied to lead. The biokinetic models of plutonium, americium, and neptunium were updated based on a generic model for the actinides. Ingestion dose coefficients (committed equivalent doses to tissues and organs and committed effective doses) were given for selected radioisotopes of these elements and for the elements in *ICRP Publication 56*. Account was taken of routes of urinary and faecal excretion in estimates of doses to the urinary bladder and colon. The revised values of w_T, given by the Commission in the 1990 recommendations for a wider range of tissues (ICRP, 1991), were used in the calculation of committed effective dose. The ingestion dose coefficients superseded those in *ICRP Publication 56*.

In *ICRP Publication 69* (ICRP, 1995a), age-specific biokinetic models and ingestion dose coefficients were given for radioisotopes of iron, selenium, antimony, thorium, and uranium, using the methodology adopted in *ICRP Publication 67*. The generic model for the actinides was used for thorium and that for the alkaline earths was used for uranium. A new, physiologically based, age-dependent biokinetic model for iron was also introduced.

In 1994, *ICRP Publication 66* gave a revised kinetic and dosimetric model of the Human Respiratory Tract, which is also age-specific (ICRP, 1994a). Using this new model of the respiratory tract, *ICRP Publication 71* (ICRP, 1996) gave inhalation dose coefficients for selected radioisotopes of the 29 elements covered in *ICRP Publications 56, 67* and *69*, and for isotopes of calcium and curium, for which biokinetic models were included in the report. The inhalation dose coefficients superseded those in *ICRP Publication 56*. Dose coefficients were also included in the report for important gaseous and vapour compounds of the 31 elements. Table 1 summarizes the contents of *ICRP Publications 56, 67, 69* and *71*.

The present report is a compilation of age-dependent committed effective dose coefficients for ingestion and inhalation of radionuclides of the 31 elements covered in *ICRP Publications 56, 67, 69* and *71*. Additional radioisotopes of these 31 elements have been included from those given in the earlier Publications.

Table 1. Summary of reports on age-dependent dose coefficients for members of the public from intake of radionuclides

	Publications on age-dependent dose coefficients			
ICRP Publication	56[a]	67[b]	69[c]	71[d]
Ingestion dose coefficients	+	+	+	−
Gastrointestinal tract model[e]	30	30	30	30
Inhalation dose coefficients	+	−	−	+
Respiratory tract model[e]	30	na	na	66
Tissue weighting factors[e]	26	60	60	60

[a] For selected radioisotopes of H, C, Sr, Zr, Nb, Ru, I, Cs, Ce, Pu, Am and Np.
[b] For selected radioisotopes of S, Co, Ni, Zn, Mo, Tc, Ag, Te, Ba, Pb, Po and Ra (model updates for Sr, Pu, Am and Np).
[c] For selected radioisotopes of Fe, Sb, Se, Th and U.
[d] For selected radioisotopes of elements in Parts 1, 2 and 3, plus Ca and Cm.
[e] ICRP Publication number.
+/−, dose coefficients given/not given in report.
na, not applicable.

Dose coefficients for members of the public are also given for radioisotopes of a further 60 elements, for which dose coefficients for workers were given in *ICRP Publication 68* (ICRP, 1994b). The biokinetic models for adults given in *ICRP Publication 30* (ICRP, 1979, 1980, 1981, 1988) are applied to calculate these dose coefficients, except that age-specific excretion rates from the urinary bladder are used, and increased gastrointestinal absorption in infants is assumed. Changes in body mass and tissue geometry in children are also taken into account. Because changes in biokinetics are considered with age and have not been considered fully, these additional dose coefficients should be used with care for assessing doses to infants and children (see Paragraph 17). Equivalent doses to tissues are not included in this report.

The membership of the Task Group on Dose Calculations at the time of preparation of the report was:

Members

K. F. Eckerman (Chairman)   R. W. Leggett
V. Berkovski   I. A. Likhtarev
L. Bertelli   D. Noßke
M. Cristy   A. W. Phipps
G. M. Kendall

Corresponding Members

T. Dillman   L. Johansson
K. Henrichs   A. R. Reddy
A. C. James

The membership of the Task Group on Internal Dosimetry at the time of preparation of the report was:

Members

J. W. Stather (Chairman)	J. Inaba
M. R. Bailey	R. W. Leggett
A. Bouville	H. Métivier
F. T. Cross	D. Noßke
K. F. Eckerman*	M. Roy
R. A. Guilmette	D. M. Taylor
J. D. Harrison	

Corresponding Members

J. C. Barton	G. M. Kendall
P.-G. Beau	N. Nelson
X. Chen	J. Piechowski
M. Cristy	V. Repin
F. A. Fry	M. Sikov
O. Hoffman	

During the period of preparation of this report, the membership of Committee 2 was:

A. Kaul (Chairman)	I. A. Likhtarev
A. Bouville	H. Métivier
X. Chen	H. G. Paretzke
F. T. Cross	A. R. Reddy
G. Dietze	M. Roy
K. F. Eckerman	J. W. Stather
F. A. Fry	D. M. Taylor
J. Inaba	R. H. Thomas

Acknowledgements

The work of the Task Groups was aided by significant technical contributions relating to the calculation of dose coefficients by T. P. Fell and T. J. Silk.

---

* Chairman of the Dose Calculations and Reference Man Task Groups, is an 'ex officio'; member of the Task Group.

GLOSSARY OF TERMS

Absorbed Dose

the physical dose quantity which is given by:

$$D = \frac{d\bar{\varepsilon}}{dm}$$

where $d\bar{\varepsilon}$ is the mean energy imparted by ionizing radiation to the matter in a volume element and dm is the mass of the matter in this volume element. The SI unit for absorbed dose is joule per kilogram ($J \ kg^{-1}$) and its special name is gray (Gy).

Absorbed Fraction ($AF(T \leftarrow S)_R$)

the fraction of energy emitted as a specified radiation type R in a specified source region S, which is absorbed in a specified target tissue T.

Aerodynamic Diameter (d_{ae})

diameter (μm) of a unit density (1 g cm^{-3}) sphere that has the same terminal settling velocity in air as the particle of interest.

Alveolar–Interstitial Region (AI)

consists of the respiratory bronchioles, alveolar ducts and sacs with their alveoli, and the interstitial connective tissue; airway generations 16 and beyond.

Activity Median Aerodynamic Diameter (AMAD)

of the activity in the aerosol, 50% is associated with particles of aerodynamic diameter (d_{ae}) greater than the AMAD. Used when deposition depends principally on inertial impaction and sedimentation, typically when the AMAD is greater than about 0.5 μm.

Becquerel (Bq)

the special name for the SI unit of activity, 1 Bq = 1 s^{-1}.

Bronchial Region (BB)

consists of the trachea and bronchi.

Bronchiolar Region (bb)

consists of the bronchioles and terminal bronchioles.

Cells Near Bone Surfaces

those cells within 10 μm of endosteal and bone surfaces lined with epithelium.

Clearance from the Respiratory Tract

the removal of material from the respiratory tract by particle transport and by absorption into body fluids.

Committed Effective Dose $(E(\tau))$
the sum of the products of the committed organ or tissue equivalent doses and the appropriate organ or tissue weighting factors (w_T), where τ is the integration time in years following the intake. The integration time is 50 y for adults, and from intake to age 70 y for children.

Committed Equivalent Dose $(H_T(\tau))$
the time integral of the equivalent dose rate in a particular tissue or organ that will be received by an individual following intake of radioactive material into the body, where τ is the integration time in years following the intake. The integration time is 50 y for adults, and from intake to age 70 y for children.

Deposition in the Respiratory Tract
refers to the initial processes determining how much of the material in the inspired air remains behind in the respiratory tract after exhalation. Deposition of material may occur during both inspiration and exhalation.

Deposition Classes of Gases and Vapours According to Solubility and Reactivity
Class SR-0: Insoluble and nonreactive. Negligible deposition in the respiratory tract.
Class SR-1: Soluble or reactive. Deposition throughout the respiratory tract, which may be complete or incomplete.
Class SR-2: Highly soluble or reactive. Complete deposition in the respiratory tract with instantaneous uptake to body fluids.

Dose Coefficient
committed tissue equivalent dose per unit intake at age t_0, $h_T(\tau)$, or committed effective dose per unit intake, $e(\tau)$, where τ is the time period in years over which the dose is calculated, i.e. 50 for adults and $(70 - t_0)$ for children.

Effective Dose (E)
the sum of the weighted equivalent doses in all tissues and organs of the body, given by the expression:

$$E = \sum_T w_T \, H_T$$

where H_T is the equivalent dose in tissue or organ T and w_T is the weighting factor for tissue T.

Endogenous Excretion
term used in this context to specify the excretion of materials from body fluids to the gastrointestinal (GI) tract, applying to biliary excretion and passage of materials through the GI tract wall.

Equivalent Dose (H_T)
the equivalent dose, $H_{T,R}$, in tissue or organ T due to radiation R, is given by:

$$H_{T,R} = w_R D_{T,R}$$

where $D_{T,R}$ is the average absorbed dose from radiation R in tissue T and w_R is the radiation weighting factor. Since w_R is dimensionless, the units of equivalent dose are the same as for the absorbed dose, J kg^{-1}, and its special name is sievert (Sv). The total equivalent dose, H_T, is the sum of $H_{T,R}$ over all radiation types

$$H_T = \sum_R H_{T,R}.$$

Extrathoracic (ET) Airways
consist of anterior nose (ET$_1$) and the posterior nasal passages, mouth, pharynx and larynx (ET$_2$).

Fractional Absorption in the Gastrointestinal Tract (f_1)
the f_1 value is the fraction of an element directly absorbed from the gut to body fluids.

Gray (Gy)
the special name for the SI unit of absorbed dose. 1 Gy = 1 J kg^{-1}.

Intake
activity that enters the respiratory or gastrointestinal tract from the environment.

Organ Dose
the tissue- or organ-average absorbed dose, D_T, is given by:

$$D_T = \frac{\varepsilon_T}{m_T}$$

where ε_T is the total energy imparted in a tissue or organ T, and m_T is the mass of that tissue or organ.

Radiation Weighting Factor (w_R)
a dimensionless factor used to derive the equivalent dose from the absorbed dose averaged over a tissue or organ and is based on the quality of radiation (ICRP, 1991).

Red Bone Marrow (active)
the component of marrow which contains the bulk of the haematopoietic stem cells.

Reference Man
an adult person with the anatomical and physiological characteristics defined in the report of the ICRP Task Group on Reference Man (ICRP, 1975).

Sievert (Sv)
the special name for the SI unit of equivalent dose and effective dose. 1 Sv = 1 J kg^{-1}

Thoracic (TH) Airways
combined bronchial, bronchiolar and alveolar–interstitial regions.

Tissue Weighting Factor (w_T)

the factor by which the equivalent dose in a tissue or organ is weighted to represent the relative contribution of that tissue or organ to the total detriment resulting from uniform irradiation of the body (ICRP, 1991).

Types of Materials According to their Rates of Absorption from the Respiratory Tract to Body Fluids

Type F: deposited materials that are readily absorbed into body fluids from the respiratory tract (Fast absorption).

Type M: deposited materials that have intermediate rates of absorption into body fluids from the respiratory tract (Moderate absorption).

Type S: deposited materials that are relatively insoluble in the respiratory tract (Slow absorption).

Type V: deposited materials that, for dosimetric purposes, are assumed to be instantaneously absorbed into body fluids from the respiratory tract; in this report applied only to certain gases and vapours (Very fast absorption).

Uptake

activity that enters the body fluids from the respiratory tract, gastrointestinal tract or through the skin.

1. INTRODUCTION

(1) This report is a compilation of age-dependent committed effective dose coefficients for members of the public from intakes by ingestion and inhalation of radioisotopes of the 31 elements covered by *ICRP Publications 56, 67, 69* and *71* (ICRP, 1989, 1993, 1995a, 1996). This report also gives committed effective dose coefficients for members of the public from radioisotopes of the additional 60 elements for which dose coefficients are given for workers in *ICRP Publication 68* (ICRP, 1994b). This report does not give committed equivalent doses to tissues and organs. In a few instances there are small changes in ingestion dose coefficients from those given in previous reports (see Paragraph 6). In these cases, the values given here supersede the earlier values. These ICRP dose coefficients have been adopted in the International Basic Safety Standards (IAEA, 1996) and in the Euratom Directive (EC, 1996).

(2) Values of dose coefficients for workers are also given in the International Basic Safety Standards and in the Euratom Directive. With the exception of inhalation dose coefficients for ^{226}Ra, these values are the same as those given in *ICRP Publication 68* (ICRP, 1994b).

2. INGESTION

(3) For exposure of members of the public to radionuclides, ingestion is generally the most significant route of intake. Elements incorporated into food may be more readily absorbed from the gastrointestinal (GI) tract than inorganic forms of these elements (ICRP, 1989). This is taken into account, as far as is possible, in the choice of the fractional absorption, f_1, values adopted in *ICRP Publications 56, 67, 69* and *71*. For radioisotopes of the additional 60 elements for which dose coefficients are given in *ICRP Publications 30* and *68* (ICRP, 1979, 1980, 1981, 1994b) no specific f_1 values for absorption in food have been given. The f_1 values used here for members of the public, other than newborn, are those given for a working population in *ICRP Publications 30* and *68*.

(4) Absorption of radionuclides is increased in the newborn. On the basis of available animal data, supported by limited human data, a general approach to determine f_1 values for infants was suggested by an Expert Group set up by the Nuclear Energy Agency (NEA/OECD, 1988). In general, this approach has been used in *ICRP Publications 56, 67*, and *69* (ICRP, 1989, 1993, 1995a). For fractional absorption values greater than 0.5 in the adult, complete absorption in the infant is assumed. For absorption values between 0.5 and 0.01 in the adult, an increase by a factor of two is assumed for infants. For values in the adult of 0.001 or less, an increase by a factor of ten is assumed. For palladium, beryllium, and hafnium, for which adult values are 0.005, 0.005 and 0.002, respectively, infant values of 0.05, 0.02 and 0.02, respectively, are used.

(5) Intermediate f_1 values have also been specified for a number of elements (calcium, iron, cobalt, strontium, barium, lead and radium). Table A1 gives the f_1 values adopted in *ICRP Publications 56, 67, 69*, and *71* (ICRP, 1989, 1993, 1995a, 1996) and used in this report for the calculation of ingestion dose coefficients for members of the public.

(6) There are some small changes to the ingestion dose coefficients given in this report compared with those given in *Publications 67* and *69* (ICRP, 1993, 1995a). Some result from a change in the application of the splitting rule to remainder tissues in the calculation of the effective dose. This rule states that in those cases in which the most exposed remainder tissue or organ receives the highest committed equivalent dose of all organs, a weighting factor of 0.025 (half of remainder) is applied to that tissue or organ, and 0.025 to the mass average committed equivalent dose in the rest of the remainder tissues and organs (ICRP, 1993). The application of the splitting rule is now based on the rounded (two digits as published) value of the equivalent dose to the remainder tissues, rather than the value as represented within the computer. As a result, it is not now being applied in a number of cases where many tissues receive very similar doses. The changes to effective dose are generally small. A minor error in the f_1 value for ingestion of ^{210}Po by infants has been corrected. An improvement in the treatment of activity in other tissues has been made in computations involving independent decay product kinetics, as explained in Annexe C of *ICRP Publication 71* (ICRP, 1996). There are also very small differences due to rounding in the calculations carried out at the different organizations involved (see Paragraph 27).

3

3. INHALATION

(7) Inhalation dose coefficients for members of the public have been calculated using the new model for the Human Respiratory Tract, given in *ICRP Publication 66* (ICRP, 1994a), which quantifies the retention of deposited activity in the various respiratory tract regions. Clearance from these regions is represented by three processes. It is assumed that clearance from the anterior nose (ET_1) is extrinsic (e.g. nose-blowing), and that elsewhere it results from competition between particle transport to the GI tract and lymph nodes, and absorption of material into body fluids. It is assumed, by default, that particle transport rates are the same for all materials, and that the absorption rates are the same in all regions except ET_1, where none occurs. It is recommended (ICRP, 1994a) that material-specific rates of absorption should be used whenever reliable human or animal experimental data exist. For other compounds, default values are recommended for use according to whether absorption is considered to be fast (Type F), moderate (Type M) or slow (Type S) (corresponding broadly to inhalation Classes D, W and Y in the *ICRP Publication 30* model). It is assumed that clearance rates of all three processes—nose-blowing, particle transport and absorption—are independent of age and sex. For members of the public, dose coefficients are based on an activity median aerodynamic diameter, AMAD, of 1 µm and specified distributions of time spent at four levels of exercise: sleep, sitting, light exercise and heavy exercise (ICRP, 1996).

(8) For the calculation of inhalation dose coefficients, allowance has to be made for the absorption of material passing through the GI tract after clearance from the respiratory system. In assigning f_1 values for the 31 elements in *ICRP Publication 71* (ICRP, 1996), it was considered that for environmental exposure by inhalation, the radionuclides might typically be present as minor constituents of the inhaled particles, and that therefore absorption into body fluids would depend on dissolution of the particle matrix, as well as on the elemental form of the radionuclide. Generally, for Type F materials, the greatest f_1 value for the element in *ICRP Publication 68* (ICRP, 1994b) is applied. For Types M and S, default f_1 values of 0.1 and 0.01, respectively, are applied, unless a lower f_1 value for that absorption Type (or for a more soluble Type) was used in *ICRP Publication 68*, in which case that value is applied. For the remaining 60 elements, the f_1 values adopted in *ICRP Publication 30* were used.

(9) Higher f_1 values have been adopted for 3-month-old infants for all 91 elements using the approach used for ingestion in *ICRP Publications 56, 67* and *69*. For most elements, adult values are applied to 1-, 5-, 10- and 15-y-old children. However, for Type F forms of calcium, iron, cobalt, strontium, barium, lead and radium, intermediate values are used for children following the approach in *ICRP Publications 56, 67, 69* and *71*. For Types M and S forms of these elements, the default adult f_1 values (0.1 and 0.01, respectively) are also applied to children.

(10) In the case of the 31 elements for which information on lung absorption is given in ICRP *Publication 71* (ICRP, 1996), dose coefficients are given for the three absorption Types, together with a recommended default when no specific information is available on the chemical form of the radionuclide. The default Types are given in Table 2.

(11) The inhalation dose coefficients for the various radionuclides of the additional 60 elements have been calculated on the basis that compounds assigned to lung inhalation Classes D, W and Y in *ICRP Publication 30* (Parts 1–4) have been assigned to absorption Types F, M and S, respectively, as in *ICRP Publication 68* (1994b). Information is given in

Table 2. Biokinetic data and models used to calculate committed effective dose per unit intake via inhalation for exposure to particulate aerosols or to gases and vapours for members of the public

Element	Lung absorption Type(s)[a]	Classes for gases/vapours[c]	ICRP for details of biokinetic model[d] and absorption Type(s)
Hydrogen	F, M[b], S	SR-1, SR-2[d]	56, 67 and 71
Beryllium	M, S		30, Part 3
Carbon	F, M[b], S	SR-1, SR-2[e]	56, 67 and 71
Fluorine	F, M, S		30, Part 2
Sodium	F		30, Part 2
Magnesium	F, M		30, Part 3
Aluminium	F, M		30, Part 3
Silicon	F, M, S		30, Part 3
Phosphorus	F, M		30, Part 1
Sulphur	F, M[b], S	SR-1	67 and 71
Chlorine	F, M		30, Part 2
Potassium	F		30, Part 2
Calcium	F, M[b], S		71
Scandium	S		30, Part 3
Titanium	F, M, S		30, Part 3
Vanadium	F, M		30, Part 3
Chromium	F, M, S		30, Part 2
Manganese	F, M		30, Part 1
Iron	F, M[b], S		69 and 71
Cobalt	F, M[b], S		67 and 71
Nickel	F, M[b], S	SR-1	67 and 71
Copper	F, M, S		30, Part 2
Zinc	F, M[b], S		67 and 71
Gallium	F, M		30, Part 3
Germanium	F, M		30, Part 3
Arsenic	M		30, Part 3
Selenium	F[b], M, S		69 and 71
Bromine	F, M		30, Part 2
Rubidium	F		30, Part 2
Strontium	F, M[b], S		67 and 71
Yttrium	M, S		30, Part 2
Zirconium	F, M[b], S		56, 67 and 71
Niobium	F, M[b], S		56, 67 and 71
Molybdenum	F, M[b], S		67 and 71
Technetium	F, M[b], S		67 and 71
Ruthenium	F, M[b], S	SR-1	56, 67 and 71
Rhodium	F, M, S		30, Part 2
Palladium	F, M, S		30, Part 3
Silver	F, M[b], S		67 and 71
Cadmium	F, M, S		30, Part 2
Indium	F, M		30, Part 2
Tin	F, M		30, Part 3
Antimony	F, M[b], S		69 and 71
Tellurium	F, M[b], S	SR-1	67 and 71
Iodine	F[b], M, S	SR-1	56, 67 and 71
Caesium	F[b], M, S		56, 67 and 71
Barium	F, M[b], S		67 and 71
Lanthanum	F, M		30, Part 3
Cerium	F, M[b], S		56, 67 and 71
Praseodymium	M, S		30, Part 3
Neodymium	M, S		30, Part 3
Promethium	M, S		30, Part 3
Samarium	M		30, Part 3
Europium	M		30, Part 3
Gadolinium	F, M		30, Part 3
Terbium	M		30, Part 3
Dysprosium	M		30, Part 3

Table 2. —(continued)

Element	Lung absorption Type(s)[a]	Classes for gases/vapours[cc]	ICRP for details of biokinetic model[d] and absorption Type(s)
Holmium	M		30, Part 3
Erbium	M		30, Part 3
Thulium	M		30, Part 3
Ytterbium	M, S		30, Part 3
Lutetium	M, S		30, Part 3
Hafnium	F, M		30, Part 3
Tantalum	M, S		30, Part 3
Tungsten	F		30, Part 3
Rhenium	F, M		30, Part 2
Osmium	F, M, S		30, Part 2
Iridium	F, M, S		30, Part 2
Platinum	F		30, Part 3
Gold	F, M, S		30, Part 2
Mercury	F, M	SR-1	30, Part 2
Thallium	F		30, Part 3
Lead	F, M[b], S		67 and 71
Bismuth	F, M		30, Part 2
Polonium	F, M[b], S		67 and 71
Astatine	F, M		30, Part 3
Francium	F		30, Part 3
Radium	F, M[b], S		67 and 71
Actinium	F, M, S		30, Part 3
Thorium	F, M, S[b]		69 and 71
Protactinium	M, S		30, Part 3
Uranium	F, M[b], S		69 and 71
Neptunium	F, M[b], S		69 and 71
Plutonium	F, M[b], S		67 and 71
Americium	F, M[b], S		67 and 71
Curium	F, M[b], S		71
Berkelium	M		30, Part 4
Californium	M		30, Part 4
Einsteinium	M		30, Part 4
Fermium	M		30, Part 4
Mendelevium	M		30, Part 4

[a]For particulates: F, fast; M, moderate; S, slow. [b]Recommended default absorption Type for particulate aerosol when no specific information is available (see *ICRP Publication 71*, 1996). [c]Also for ingestion dose coefficients. [d]Tritium gas and tritiated methane, class SR-1; tritiated water and unspecified compounds, class SR-2. [e]Carbon monoxide, class SR-1; carbon dioxide and organic compounds, class SR-2.

the relevant ICRP Publications (see Table 2) on the chemical forms appropriate to the different inhalation Classes/Types.

(12) The lung absorption Types and corresponding f_1 values adopted for the calculation of inhalation dose coefficients are included in Table A.2.

(13) For radionuclides inhaled in particulate form, it is assumed that entry and regional deposition in the respiratory tract are governed only by physical properties of the aerosol, such as the size distribution of the particles. The situation is different for gases and vapours, for which respiratory tract deposition is material-specific. Almost all inhaled gas molecules contact airway surfaces, but usually return to the air unless they dissolve in, or react with, the surface lining. The fraction of an inhaled gas or vapour that is deposited in each region thus depends on its solubility and reactivity.

(14) As a general default approach, the *ICRP Publication 66* model assigns gases and vapours to three classes, on the basis of the initial pattern of respiratory tract deposition:

- Class SR-0 insoluble and non-reactive: negligible deposition in the respiratory tract.
- Class SR-1 soluble or reactive: deposition may occur throughout the respiratory tract.
- Class SR-2 highly soluble or reactive: total deposition in the extrathoracic airways (ET_2).

Subsequent retention in the respiratory tract and absorption to body fluids are determined by the chemical properties of the specific gas or vapour. *ICRP Publications 68* and *71* give information on the assignment of gases and vapours to these three classes, and for selected Class SR-1 compounds information on fractional deposition and subsequent clearance. For Class SR-0 gases considered here, which are all noble gases, Table A.4 gives effective dose rates from submersion in a cloud of the gas. Both external irradiation from the cloud, and internal irradiation from the gas within the respiratory tract are included, but for most radionuclides considered here, the former dominates.

(15) As an alternative to any of the three default Types defined in *ICRP Publication 66*, very fast uptake to body fluids (Type V) may be recommended. Although consideration has to be given to the total respiratory tract deposition, regional deposition does not need to be assessed for such materials, since, for the purposes of dose calculation, they can be treated as if they were injected directly into body fluids.

4. SYSTEMIC ACTIVITY

(16) The age-specific biokinetic models given in *ICRP Publications 56, 67, 69* and *71* for 31 elements are used to describe the tissue distribution, retention and excretion of systemic activity.

(17) For radioisotopes of the remaining 60 elements covered by the present report, the biokinetic models used are based on those given in *Publication 30* (Parts 1–4) for workers. The dose calculations for the radionuclides of these additional elements allow for age-specific changes in gut uptake; body mass and geometry; and excretion rates from the urinary bladder, but not in the biokinetics of systemic activity. The biokinetic models have also been extended to give excretion pathways. Details of the parameters for urinary and faecal excretion are given in Table 3. The coefficients are therefore based on biokinetic data that have not been subject to recent review, and therefore serve only as a guide. In general it is considered that the use of adult biokinetic parameters in the calculation of doses for younger ages will lead to overestimates of the dose coefficients because of the faster rate of elimination of many elements from the body compared with adults.

(18) In the 1990 Recommendations (ICRP, 1991), the urinary bladder and colon are given explicit w_T values. Consequently, a urinary bladder model was developed for calculating the equivalent dose to the bladder wall from activity in the urine, as described in *ICRP Publications 67* and *69*.

(19) Some of the radionuclides considered here decay to nuclides which are themselves radioactive. For most elements, the treatment of decay products follows that in *ICRP Publication 30* (ICRP, 1979, 1980, 1981). The usual assumption is that decay products produced *in vivo* adopt the absorption parameters of their parent if they are produced in the respiratory or gastrointestinal tract, and the biokinetics of their parent if they are produced after absorption to blood. However, if the decay products are isotopes of noble gases, it is assumed that a fraction of the noble gas activity leaves the body instantaneously, the numerical value of that fraction depending on the half-life and site of production. For intakes of lead, radium, tellurium, thorium and uranium, separate biokinetics for systemic activity have been applied to the parent and its decay products, as decribed in Annexe C of *ICRP Publication 71* (ICRP, 1996). In all cases, the dose coefficients corresponding to the intakes of the parent radionuclide include contributions from parents and their decay products.

(20) The sources of the biokinetic models used in the calculation of ingestion and inhalation dose coefficients are given in Table 2.

Table 3. Excretion pathways adopted for systemic activity[e]

Element	Urinary to faecal excretion ratio Ratio	ICRP Publication
H	a	67
C	a	67
Na	100	10A
P	9	54
S	9	67
Cl	100	10A
Ca	1	10
Fe	b	69
Co	6	67
Ni	20	67
Zn	0.25	67
Ge	c	30
Se	2	69
Rb	3	10
Sr	b	67
Zr	5	67
Nb	5	67
Mo	8	67
Tc	1	67
Ru	4	67
Ag	0.05	67
Sb	4	69
Te	4	67
I	b	67
Cs	4	67
Ba	b	67
Ce	0.11	67
Au	d	10
Pb	b	67
Po	0.5	67
Ra	b	67
Th	b	69
U	b	69
Np	b	67
Pu	b	67
Am	b	67

[a]The excretion pathways are not considered in the recommended biokinetic parameters; bladder wall and upper and lower large intestine are assumed to receive the same dose as other tissues.

[b]The excretion pathways are considered explicitly by the recommended biokinetic model.

[c]Excretion only in urine for activity deposited in kidney. A ratio of 1 is assumed for activity in other tissues.

[d]Excretion only in urine assumed.

[e]For other elements, a ratio of 1 is assumed.

5. DOSE COEFFICIENTS FOR INGESTION AND INHALATION

(21) Details of the computational method used in the calculation of equivalent doses to tissues and committed effective doses are given in *ICRP Publications 67, 69* and *71* (ICRP, 1993, 1995a, 1996).

(22) Dose coefficients for intakes of radionuclides by 3-mo-old infants, 1-, 5-, 10-, and 15-y-old children, and adults are given in Annexe A. Ingestion dose coefficients are given in Table A.1, and inhalation dose coefficients for intakes of particulate aerosols in Table A.2. Age-dependent dose coefficients for inhalation of soluble or reactive gases are given in Table A.3, and effective dose rates for exposure of adult members of the public to inert gases are given in Table A.4. Dose rates for children have not been calculated.

(23) In most cases the adult is taken to be aged 20 y. Exceptions are made for calcium, strontium, barium, radium, lead, thorium, uranium, neptunium, plutonium, americium and curium (ICRP, 1993, 1995a, 1996). For these elements, adults are taken to be aged 25 y (ICRP, 1995b).

(24) In the calculations of the activity in source regions of the body following intakes at the specified ages, continuous changes with age in the transfer rates governing distribution and retention are obtained by linear interpolation according to age. This also applies to the transfer of activity from the gastrointestinal tract. For application to other ages and to protracted intakes, the Task Group considered that doses can be estimated by applying the age-specific dose coefficients to the age ranges given below:

3 mo:	from 0 to 1 y of age
1 y:	from 1 y to 2 y
5 y:	more than 2 y to 7 y
10 y:	more than 7 y to 12 y
15 y:	more than 12 y to 17 y
Adult:	more than 17 y

As in *ICRP Publications 56, 67, 69* and *71*, a single reference subject is used to calculate dose coefficients for each age-group. Generally, parameter values for males have been adopted because of the availability of biokinetic data. Where there are known differences between sexes in the biokinetics of an element, this is noted in the relevant section of the biokinetic data in *ICRP Publication 56, 67, 69* or *71*.

(25) The tissue weighting factors, w_T, and radiation weighting factors, w_R, used for calculating the effective dose coefficients are those given by ICRP in its 1990 recommendations (ICRP, 1991).

(26) The dose coefficients calculated in this report, as in *ICRP Publications 56, 67, 69* and *71*, are for acute intakes. For intakes over extended periods, effective doses could be somewhat lower than those calculated here, where growth is significant during the period of intake. However, since the age ranges were selected to account for significant changes in growth and biokinetics during life, these coefficients can also be applied to chronic intakes for protection purposes by determining the committed dose for each year's intake and summing for intakes over all years.

(27) The new biokinetic models, which are based on the underlying physiological processes, offer a number of advantages, particularly with regard to developing age-dependent parameters. They are, however, considerably more complex than those given in

ICRP Publication 30 and more difficult to implement. The Commission has, therefore, attached particular importance to the question of Quality Assurance. The Task Group on Dose Calculations has arranged for the dose coefficients to be calculated independently, using different computer codes, in three of the contributing laboratories. Any discrepancies in these calculations exceeding those caused by rounding errors have been investigated and resolved before the publication of results.

References

EC (1996) *Council Directive 96/29/EURATOM of 13 May 1996, Laying Down the Basic Safety Standards for the Protection of the Health of Workers and the General Public Against the Dangers Arising from Ionizing Radiation.*

IAEA (1996) *International Basic Safety Standards for Protection against Ionising Radiation and for the Safety of Radioactive Sources.* Jointly sponsored by FAO, IAEA, ILO, OECD/NEA, PAHO, WHO, IAEA Safety Series No. 115. International Atomic Energy Agency, Vienna.

ICRP (1975) *Report of the Task Group on Reference Man.* ICRP Publication 23, Pergamon Press, Oxford.

ICRP (1977) *Recommendations of the International Commission on Radiological Protection.* ICRP Publication 26 *Annals of the ICRP* **1**(3), Pergamon Press, Oxford. Reprinted (with additions) in 1987.

ICRP (1979) *Limits for Intakes of Radionuclides by Workers.* ICRP Publication 30, Part 1. *Annals of the ICRP* **2**(3/4), Pergamon Press, Oxford.

ICRP (1980) *Limits for Intakes of Radionuclides by Workers.* ICRP Publication 30, Part 2. *Annals of the ICRP* **4**(3/4), Pergamon Press, Oxford.

ICRP (1981) *Limits for Intakes of Radionuclides by Workers.* ICRP Publication 30, Part 3 (including addendum to Parts 1 and 2). *Annals of the ICRP* **6**(2/3), Pergamon Press, Oxford.

ICRP (1988) *Limits for Intakes of Radionuclides by Workers: An Addendum.* ICRP Publication 30. Part 4. *Annals of the ICRP* **19**(4), Pergamon Press, Oxford.

ICRP (1989) *Age-dependent Doses to Members of the Public from Intake of Radionuclides: Part 1.* ICRP Publication 56. *Annals of the ICRP*, **20**(2), Pergamon Press, Oxford.

ICRP (1991) *1990 Recommendations of the International Commission on Radiological Protection.* ICRP Publication 60. *Annals of the ICRP* **21**(1–3), Pergamon Press, Oxford.

ICRP (1993) *Age-dependent Doses to Members of the Public from Intake of Radionuclides: Part 2, Ingestion Dose Coefficients.* ICRP Publication 67. *Annals of the ICRP* **23**(3/4), Elsevier Science Ltd, Oxford.

ICRP (1994a) *Human Respiratory Tract Model for Radiological Protection.* ICRP Publication 66. *Annals of the ICRP* **24**(1–3), Elsevier Science Ltd, Oxford.

ICRP (1994b) *Dose Coefficients for Intakes of Radionuclides by Workers.* ICRP Publication 68. *Annals of the ICRP* **24**(4), Elsevier Science Ltd, Oxford.

ICRP (1995a) *Age-dependent Doses to Members of the Public from Intake of Radionuclides: Part 3, Ingestion Dose Coefficients.* ICRP Publication 69. *Annals of the ICRP* **25**(1), Elsevier Science Ltd, Oxford.

ICRP (1995b) *Basic Anatomical and Physiological Data for use in Radiological Protection: The Skeleton.* ICRP Publication 70. *Annals of the ICRP* **25**(2), Elsevier Science Ltd, Oxford.

ICRP (1996) *Age-dependent Doses to Members of the Public from Intake of Radionuclides: Part 4, Inhalation Dose Coefficients.* ICRP Publication 71. *Annals of the ICRP* **25**(3/4), Elsevier Science Ltd, Oxford.

NEA/OECD (1988). Committee on radiation protection and public health. *Report of an Expert Group on Gut Transfer Factors.* NEA/OECD Report, Paris.

ANNEXE A. DOSE COEFFICIENTS FOR INGESTION AND INHALATION OF RADIONUCLIDES AND EFFECTIVE DOSE RATES FOR EXPOSURE TO INERT GASES

Table A.1. Ingestion dose coefficients, $e(\tau)$, to age 70 y (Sv Bq^{-1})

		f_1	$e(\tau)$	f_1	$e(\tau)$				
Nuclide	Physical half-life	<1y	3 months	≥1y	1 Year	5 Years	10 Years	15 Years	Adult
Hydrogen[a]									
Tritiated Water	12.3 y	1.000	6.4E-11	1.000	4.8E-11	3.1E-11	2.3E-11	1.8E-11	1.8E-11
OBT	12.3 y	1.000	1.2E-10	1.000	1.2E-10	7.3E-11	5.7E-11	4.2E-11	4.2E-11
Beryllium									
Be-7	53.3 d	0.020	1.8E-10	0.005	1.3E-10	7.7E-11	5.3E-11	3.5E-11	2.8E-11
Be-10	1.60E+06 y	0.020	1.4E-08	0.005	8.0E-09	4.1E-09	2.4E-09	1.4E-09	1.1E-09
Carbon[a]									
C-11	0.340 h	1.000	2.6E-10	1.000	1.5E-10	7.3E-11	4.3E-11	3.0E-11	2.4E-11
C-14	5.73E+03 y	1.000	1.4E-09	1.000	1.6E-09	9.9E-10	8.0E-10	5.7E-10	5.8E-10
Fluorine									
F-18	1.83 h	1.000	5.2E-10	1.000	3.0E-10	1.5E-10	9.1E-11	6.2E-11	4.9E-11
Sodium									
Na-22	2.60 y	1.000	2.1E-08	1.000	1.5E-08	8.4E-09	5.5E-09	3.7E-09	3.2E-09
Na-24	15.0 h	1.000	3.5E-09	1.000	2.3E-09	1.2E-09	7.7E-10	5.2E-10	4.3E-10
Magnesium									
Mg-28	20.9 h	1.000	1.2E-08	0.500	1.4E-08	7.4E-09	4.5E-09	2.7E-09	2.2E-09
Aluminium									
Al-26	7.16E+05 y	0.020	3.4E-08	0.010	2.1E-08	1.1E-08	7.1E-09	4.3E-09	3.5E-09
Silicon									
Si-31	2.62 h	0.020	1.9E-09	0.010	1.0E-09	5.1E-10	3.0E-10	1.8E-10	1.6E-10
Si-32	4.50E+02 y	0.020	7.3E-09	0.010	4.1E-09	2.0E-09	1.2E-09	7.0E-10	5.6E-10
Phosphorus									
P-32	14.3 d	1.000	3.1E-08	0.800	1.9E-08	9.4E-09	5.3E-09	3.1E-09	2.4E-09
P-33	25.4 d	1.000	2.7E-09	0.800	1.8E-09	9.1E-10	5.3E-10	3.1E-10	2.4E-10
Sulphur[a]									
S-35 (inorganic)	87.4 d	1.000	1.3E-09	1.000	8.7E-10	4.4E-10	2.7E-10	1.6E-10	1.3E-10
S-35 (organic)	87.4 d	1.000	7.7E-09	1.000	5.4E-09	2.7E-09	1.6E-09	9.5E-10	7.7E-10

a Dose coefficients for radiosotopes of this element are based on age-specific biokinetic data

AGE-DEPENDENT DOSES FROM INTAKE OF RADIONUCLIDES

Table A.1.—(continued)

Nuclide	Physical half-life	f_1 <1y	$e(\tau)$ 3 months	f_1 ≥1y	$e(\tau)$ 1 Year	5 Years	10 Years	15 Years	Adult
Chlorine									
Cl-36	3.01E+05 y	1.000	9.8E-09	1.000	6.3E-09	3.2E-09	1.9E-09	1.2E-09	9.3E-10
Cl-38	0.620 h	1.000	1.4E-09	1.000	7.7E-10	3.8E-10	2.2E-10	1.5E-10	1.2E-10
Cl-39	0.927 h	1.000	9.7E-10	1.000	5.5E-10	2.7E-10	1.6E-10	1.1E-10	8.5E-11
Potassium									
K-40	1.28E+09 y	1.000	6.2E-08	1.000	4.2E-08	2.1E-08	1.3E-08	7.6E-09	6.2E-09
K-42	12.4 h	1.000	5.1E-09	1.000	3.0E-09	1.5E-09	8.6E-10	5.4E-10	4.3E-10
K-43	22.6 h	1.000	2.3E-09	1.000	1.4E-09	7.6E-10	4.7E-10	3.0E-10	2.5E-10
K-44	0.369 h	1.000	1.0E-09	1.000	5.5E-10	2.7E-10	1.6E-10	1.1E-10	8.4E-11
K-45	0.333 h	1.000	6.2E-10	1.000	3.5E-10	1.7E-10	9.9E-11	6.8E-11	5.4E-11
Calcium[a,b]									
Ca-41	1.40E+05 y	0.600	1.2E-09	0.300	5.2E-10	3.9E-10	4.8E-10	5.0E-10	1.9E-10
Ca-45	163 d	0.600	1.1E-08	0.300	4.9E-09	2.6E-09	1.8E-09	1.3E-09	7.1E-10
Ca-47	4.53 d	0.600	1.3E-08	0.300	9.3E-09	4.9E-09	3.0E-09	1.8E-09	1.6E-09
Scandium									
Sc-43	3.89 h	0.001	1.8E-09	1.0E-04	1.2E-09	6.1E-10	3.7E-10	2.3E-10	1.9E-10
Sc-44	3.93 h	0.001	3.5E-09	1.0E-04	2.2E-09	1.2E-09	7.1E-10	4.4E-10	3.5E-10
Sc-44m	2.44 d	0.001	2.4E-08	1.0E-04	1.6E-08	8.3E-09	5.1E-09	3.1E-09	2.4E-09
Sc-46	83.8 d	0.001	1.1E-08	1.0E-04	7.9E-09	4.4E-09	2.9E-09	1.8E-09	1.5E-09
Sc-47	3.35 d	0.001	6.1E-09	1.0E-04	3.9E-09	2.0E-09	1.2E-09	6.8E-10	5.4E-10
Sc-48	1.82 d	0.001	1.3E-08	1.0E-04	9.3E-09	5.1E-09	3.3E-09	2.1E-09	1.7E-09
Sc-49	0.956 h	0.001	1.0E-09	1.0E-04	5.7E-10	2.8E-10	1.6E-10	1.0E-10	8.2E-11
Titanium									
Ti-44	47.3 y	0.020	5.5E-08	0.010	3.1E-08	1.7E-08	1.1E-08	6.9E-09	5.8E-09
Ti-45	3.08 h	0.020	1.6E-09	0.010	9.8E-10	5.0E-10	3.1E-10	1.9E-10	1.5E-10
Vanadium									
V-47	0.543 h	0.020	7.3E-10	0.010	4.1E-10	2.0E-10	1.2E-10	8.0E-11	6.3E-11
V-48	16.2 d	0.020	1.5E-08	0.010	1.1E-08	5.9E-09	3.9E-09	2.5E-09	2.0E-09
V-49	330 d	0.020	2.2E-10	0.010	1.4E-10	6.9E-11	4.0E-11	2.3E-11	1.8E-11

a Dose coefficients for this element are based on age-specific biokinetic data

b The f_1 value for 1 to 15 year olds is 0.4

Table A.1.—(continued)

Nuclide	Physical half-life	f_1 <1y	$e(\tau)$ 3 months	f_1 ≥1y	$e(\tau)$ 1 Year	5 Years	10 Years	15 Years	Adult
Chromium									
Cr-48	23.0 h	0.200	1.4E-09	0.100	9.9E-10	5.7E-10	3.8E-10	2.5E-10	2.0E-10
		0.020	1.4E-09	0.010	9.9E-10	5.7E-10	3.8E-10	2.5E-10	2.0E·10
Cr-49	0.702 h	0.200	6.8E-10	0.100	3.9E-10	2.0E-10	1.1E-10	7.7E-11	6.1E-11
		0.020	6.8E-10	0.010	3.9E-10	2.0E-10	1.1E-10	7.7E-11	6.1E-11
Cr-51	27.7 d	0.200	3.5E-10	0.100	2.3E-10	1.2E-10	7.8E-11	4.8E-11	3.8E-11
		0.020	3.3E-10	0.010	2.2E-10	1.2E-10	7.5E-11	4.6E-11	3.7E-11
Manganese									
Mn-51	0.770 h	0.200	1.1E-09	0.100	6.1E-10	3.0E-10	1.8E-10	1.2E-10	9.3E-11
Mn-52	5.59 d	0.200	1.2E-08	0.100	8.8E-09	5.1E-09	3.4E-09	2.2E-09	1.8E-09
Mn-52m	0.352 h	0.200	7.8E-10	0.100	4.4E-10	2.2E-10	1.3E-10	8.8E-11	6.9E-11
Mn-53	3.70E+06 y	0.200	4.1E-10	0.100	2.2E-10	1.1E-10	6.5E-11	3.7E-11	3.0E-11
Mn-54	312 d	0.200	5.4E-09	0.100	3.1E-09	1.9E-09	1.3E-09	8.7E-10	7.1E-10
Mn-56	2.58 h	0.200	2.7E-09	0.100	1.7E-09	8.5E-10	5.1E-10	3.2E-10	2.5E-10
Iron[a,b]									
Fe-52	8.28 h	0.600	1.3E-08	0.100	9.1E-09	4.6E-09	2.8E-09	1.7E-09	1.4E-09
Fe-55	2.70 y	0.600	7.6E-09	0.100	2.4E-09	1.7E-09	1.1E-09	7.7E-10	3.3E-10
Fe-59	44.5 d	0.600	3.9E-08	0.100	1.3E-08	7.5E-09	4.7E-09	3.1E-09	1.8E-09
Fe-60	1.00E+05 y	0.600	7.9E-07	0.100	2.7E-07	2.7E-07	2.5E-07	2.3E-07	1.1E-07
Cobalt[a,c]									
Co-55	17.5 h	0.600	6.0E-09	0.100	5.5E-09	2.9E-09	1.8E-09	1.1E-09	1.0E-09
Co-56	78.7 d	0.600	2.5E-08	0.100	1.5E-08	9.8E-09	5.8E-09	3.8E-09	2.5E-09
Co-57	271 d	0.600	2.9E-09	0.100	1.6E-09	8.9E-10	5.8E-10	3.7E-10	2.1E-10
Co-58	70.8 d	0.600	7.3E-09	0.100	4.4E-09	2.6E-09	1.7E-09	1.1E-09	7.4E-10
Co-58m	9.15 h	0.600	2.0E-10	0.100	1.5E-10	7.8E-11	4.7E-11	2.8E-11	2.4E-11
Co-60	5.27 y	0.600	5.4E-08	0.100	2.7E-08	1.7E-08	1.1E-08	7.9E-09	3.4E-09
Co-60m	0.174 h	0.600	2.2E-11	0.100	1.2E-11	5.7E-12	3.2E-12	2.2E-12	1.7E-12
Co-61	1.65 h	0.600	8.2E-10	0.100	5.1E-10	2.5E-10	1.4E-10	9.2E-11	7.4E-11
Co-62m	0.232 h	0.600	5.3E-10	0.100	3.0E-10	1.5E-10	8.7E-11	6.0E-11	4.7E-11
Nickel[a]									
Ni-56	6.10 d	0.100	5.3E-09	0.050	4.0E-09	2.3E-09	1.6E-09	1.1E-09	8.6E-10
Ni-57	1.50 d	0.100	6.8E-09	0.050	4.9E-09	2.7E-09	1.7E-09	1.1E-09	8.7E-10

a Dose coefficients for this element are based on age-specific biokinetic data

b The f_1 value for 1 to 15 year olds is 0.2

c The f_1 value for 1 to 15 year olds is 0.3

Table A.1.—(continued)

Nuclide	Physical half-life	f_1 <1y	$e(\tau)$ 3 months	f_1 ≥1y	$e(\tau)$ 1 Year	5 Years	10 Years	15 Years	Adult
Ni-59	7.50E+04 y	0.100	6.4E-10	0.050	3.4E-10	1.9E-10	1.1E-10	7.3E-11	6.3E-11
Ni-63	96.0 y	0.100	1.6E-09	0.050	8.4E-10	4.6E-10	2.8E-10	1.8E-10	1.5E-10
Ni-65	2.52 h	0.100	2.1E-09	0.050	1.3E-09	6.3E-10	3.8E-10	2.3E-10	1.8E-10
Ni-66	2.27 d	0.100	3.3E-08	0.050	2.2E-08	1.1E-08	6.6E-09	3.7E-09	3.0E-09
Copper									
Cu-60	0.387 h	1.000	7.0E-10	0.500	4.2E-10	2.2E-10	1.3E-10	8.9E-11	7.0E-11
Cu-61	3.41 h	1.000	7.1E-10	0.500	7.5E-10	3.9E-10	2.3E-10	1.5E-10	1.2E-10
Cu-64	12.7 h	1.000	5.2E-10	0.500	8.3E-10	4.2E-10	2.5E-10	1.5E-10	1.2E-10
Cu-67	2.58 d	1.000	2.1E-09	0.500	2.4E-09	1.2E-09	7.2E-10	4.2E-10	3.4E-10
Zinc[a]									
Zn-62	9.26 h	1.000	4.2E-09	0.500	6.5E-09	3.3E-09	2.0E-09	1.2E-09	9.4E-10
Zn-63	0.635 h	1.000	8.7E-10	0.500	5.2E-10	2.6E-10	1.5E-10	1.0E-10	7.9E-11
Zn-65	244 d	1.000	3.6E-08	0.500	1.6E-08	9.7E-09	6.4E-09	4.5E-09	3.9E-09
Zn-69	0.950 h	1.000	3.5E-10	0.500	2.2E-10	1.1E-10	6.0E-11	3.9E-11	3.1E-11
Zn-69m	13.8 h	1.000	1.3E-09	0.500	2.3E-09	1.2E-09	7.0E-10	4.1E-10	3.3E-10
Zn-71m	3.92 h	1.000	1.4E-09	0.500	1.5E-09	7.8E-10	4.8E-10	3.0E-10	2.4E-10
Zn-72	1.94 d	1.000	8.7E-09	0.500	8.6E-09	4.5E-09	2.8E-09	1.7E-09	1.4E-09
Gallium									
Ga-65	0.253 h	0.010	4.3E-10	0.001	2.4E-10	1.2E-10	6.9E-11	4.7E-11	3.7E-11
Ga-66	9.40 h	0.010	1.2E-08	0.001	7.9E-09	4.0E-09	2.5E-09	1.5E-09	1.2E-09
Ga-67	3.26 d	0.010	1.8E-09	0.001	1.2E-09	6.4E-10	4.0E-10	2.4E-10	1.9E-10
Ga-68	1.13 h	0.010	1.2E-09	0.001	6.7E-10	3.4E-10	2.0E-10	1.3E-10	1.0E-10
Ga-70	0.353 h	0.010	3.9E-10	0.001	2.2E-10	1.0E-10	5.9E-11	4.0E-11	3.1E-11
Ga-72	14.1 h	0.010	1.0E-08	0.001	6.8E-09	3.6E-09	2.2E-09	1.4E-09	1.1E-09
Ga-73	4.91 h	0.010	3.0E-09	0.001	1.9E-09	9.3E-10	5.5E-10	3.3E-10	2.6E-10
Germanium									
Ge-66	2.27 h	1.000	8.3E-10	1.000	5.3E-10	2.9E-10	1.9E-10	1.3E-10	1.0E-10
Ge-67	0.312 h	1.000	7.7E-10	1.000	4.2E-10	2.1E-10	1.2E-10	8.2E-11	6.5E-11
Ge-68	288 d	1.000	1.2E-08	1.000	8.0E-09	4.2E-09	2.6E-09	1.6E-09	1.3E-09
Ge-69	1.63 d	1.000	2.0E-09	1.000	1.3E-09	7.1E-10	4.6E-10	3.0E-10	2.4E-10

a Dose coefficients for this element are based on age-specific biokinetic data

Table A.1.—(continued)

Nuclide	Physical half-life	f_1 <1y	$e(\tau)$ 3 months	f_1 ≥1y	$e(\tau)$ 1 Year	5 Years	10 Years	15 Years	Adult
Ge-71	11.8 d	1.000	1.2E-10	1.000	7.8E-11	4.0E-11	2.4E-11	1.5E-11	1.2E-11
Ge-75	1.38 h	1.000	5.5E-10	1.000	3.1E-10	1.5E-10	8.7E-11	5.9E-11	4.6E-11
Ge-77	11.3 h	1.000	3.0E-09	1.000	1.8E-09	9.9E-10	6.2E-10	4.1E-10	3.3E-10
Ge-78	1.45 h	1.000	1.2E-09	1.000	7.0E-10	3.6E-10	2.2E-10	1.5E-10	1.2E-10
Arsenic									
As-69	0.253 h	1.000	6.6E-10	0.500	3.7E-10	1.8E-10	1.1E-10	7.2E-11	5.7E-11
As-70	0.876 h	1.000	1.2E-09	0.500	7.8E-10	4.1E-10	2.5E-10	1.7E-10	1.3E-10
As-71	2.70 d	1.000	2.8E-09	0.500	2.8E-09	1.5E-09	9.3E-10	5.7E-10	4.6E-10
As-72	1.08 d	1.000	1.1E-08	0.500	1.2E-08	6.3E-09	3.8E-09	2.3E-09	1.8E-09
As-73	80.3 d	1.000	2.6E-09	0.500	1.9E-09	9.3E-10	5.6E-10	3.2E-10	2.6E-10
As-74	17.8 d	1.000	1.0E-08	0.500	8.2E-09	4.3E-09	2.6E-09	1.6E-09	1.3E-09
As-76	1.10 d	1.000	1.0E-08	0.500	1.1E-08	5.8E-09	3.4E-09	2.0E-09	1.6E-09
As-77	1.62 d	1.000	2.7E-09	0.500	2.9E-09	1.5E-09	8.7E-10	5.0E-10	4.0E-10
As-78	1.51 h	1.000	2.0E-09	0.500	1.4E-09	7.0E-10	4.1E-10	2.7E-10	2.1E-10
Selenium*									
Se-70	0.683 h	1.000	1.0E-09	0.800	7.1E-10	3.6E-10	2.2E-10	1.5E-10	1.2E-10
Se-73	7.15 h	1.000	1.6E-09	0.800	1.4E-09	7.4E-10	4.8E-10	2.5E-10	2.1E-10
Se-73m	0.650 h	1.000	2.6E-10	0.800	1.8E-10	9.5E-11	5.9E-11	3.5E-11	2.8E-11
Se-75	120 d	1.000	2.0E-08	0.800	1.3E-08	8.3E-09	6.0E-09	3.1E-09	2.6E-09
Se-79	6.50E+04 y	1.000	4.1E-08	0.800	2.8E-08	1.9E-08	1.4E-08	4.1E-09	2.9E-09
Se-81	0.308 h	1.000	3.4E-10	0.800	1.9E-10	9.0E-11	5.1E-11	3.4E-11	2.7E-11
Se-81m	0.954 h	1.000	6.0E-10	0.800	3.7E-10	1.8E-10	1.1E-10	6.7E-11	5.3E-11
Se-83	0.375 h	1.000	4.6E-10	0.800	2.9E-10	1.5E-10	8.7E-11	5.9E-11	4.7E-11
Bromine									
Br-74	0.422 h	1.000	9.0E-10	1.000	5.2E-10	2.6E-10	1.5E-10	1.1E-10	8.4E-11
Br-74m	0.691 h	1.000	1.5E-09	1.000	8.5E-10	4.3E-10	2.5E-10	1.7E-10	1.4E-10
Br-75	1.63 h	1.000	8.5E-10	1.000	4.9E-10	2.5E-10	1.5E-10	9.9E-11	7.9E-11
Br-76	16.2 h	1.000	4.2E-09	1.000	2.7E-09	1.4E-09	8.7E-10	5.6E-10	4.6E-10
Br-77	2.33 d	1.000	6.3E-10	1.000	4.4E-10	2.5E-10	1.7E-10	1.1E-10	9.6E-11
Br-80	0.290 h	1.000	3.9E-10	1.000	2.1E-10	1.0E-10	5.8E-11	3.9E-11	3.1E-11

a Dose coefficients for this element are based on age-specific biokinetic data

Table A.1.—(continued)

Nuclide	Physical half-life	f_1 <1y	$e(\tau)$ 3 months	f_1 ≥1y	$e(\tau)$ 1 Year	5 Years	10 Years	15 Years	Adult
Br-80m	4.42 h	1.000	1.4E-09	1.000	8.0E-10	3.9E-10	2.3E-10	1.4E-10	1.1E-10
Br-82	1.47 d	1.000	3.7E-09	1.000	2.6E-09	1.5E-09	9.5E-10	6.4E-10	5.4E-10
Br-83	2.39 h	1.000	5.3E-10	1.000	3.0E-10	1.4E-10	8.3E-11	5.5E-11	4.3E-11
Br-84	0.530 h	1.000	1.0E-09	1.000	5.8E-10	2.8E-10	1.6E-10	1.1E-10	8.8E-11
Rubidium									
Rb-79	0.382 h	1.000	5.7E-10	1.000	3.2E-10	1.6E-10	9.2E-11	6.3E-11	5.0E-11
Rb-81	4.58 h	1.000	5.4E-10	1.000	3.2E-10	1.6E-10	1.0E-10	6.7E-11	5.4E-11
Rb-81m	0.533 h	1.000	1.1E-10	1.000	6.2E-11	3.1E-11	1.8E-11	1.2E-11	9.7E-12
Rb-82m	6.20 h	1.000	8.7E-10	1.000	5.9E-10	3.4E-10	2.2E-10	1.5E-10	1.3E-10
Rb-83	86.2 d	1.000	1.1E-08	1.000	8.4E-09	4.9E-09	3.2E-09	2.2E-09	1.9E-09
Rb-84	32.8 d	1.000	2.0E-08	1.000	1.4E-08	7.9E-09	5.0E-09	3.3E-09	2.8E-09
Rb-86	18.7 d	1.000	3.1E-08	1.000	2.0E-08	9.9E-09	5.9E-09	3.5E-09	2.8E-09
Rb-87	4.70E+10 y	1.000	1.5E-08	1.000	1.0E-08	5.2E-09	3.1E-09	1.8E-09	1.5E-09
Rb-88	0.297 h	1.000	1.1E-09	1.000	6.2E-10	3.0E-10	1.7E-10	1.2E-10	9.0E-11
Rb-89	0.253 h	1.000	5.4E-10	1.000	3.0E-10	1.5E-10	8.6E-11	5.9E-11	4.7E-11
Strontium[a,b]									
Sr-80	1.67 h	0.600	3.7E-09	0.300	2.3E-09	1.1E-09	6.5E-10	4.2E-10	3.4E-10
Sr-81	0.425 h	0.600	8.4E-10	0.300	4.9E-10	2.4E-10	1.4E-10	9.6E-11	7.7E-11
Sr-82	25.0 d	0.600	7.2E-08	0.300	4.1E-08	2.1E-08	1.3E-08	8.7E-09	6.1E-09
Sr-83	1.35 d	0.600	3.4E-09	0.300	2.7E-09	1.4E-09	9.1E-10	5.7E-10	4.9E-10
Sr-85	64.8 d	0.600	7.7E-09	0.300	3.1E-09	1.7E-09	1.5E-09	1.3E-09	5.6E-10
Sr-85m	1.16 h	0.600	4.5E-11	0.300	3.0E-11	1.7E-11	1.1E-11	7.8E-12	6.1E-12
Sr-87m	2.80 h	0.600	2.4E-10	0.300	1.7E-10	9.0E-11	5.6E-11	3.6E-11	3.0E-11
Sr-89	50.5 d	0.600	3.6E-08	0.300	1.8E-08	8.9E-09	5.8E-09	4.0E-09	2.6E-09
Sr-90	29.1 y	0.600	2.3E-07	0.300	7.3E-08	4.7E-08	6.0E-08	8.0E-08	2.8E-08
Sr-91	9.50 h	0.600	5.2E-09	0.300	4.0E-09	2.1E-09	1.2E-09	7.4E-10	6.5E-10
Sr-92	2.71 h	0.600	3.4E-09	0.300	2.7E-09	1.4E-09	8.2E-10	4.8E-10	4.3E-10
Yttrium									
Y-86	14.7 h	0.001	7.6E-09	1.0E-04	5.2E-09	2.9E-09	1.9E-09	1.2E-09	9.6E-10
Y-86m	0.800 h	0.001	4.5E-10	1.0E-04	3.1E-10	1.7E-10	1.1E-10	7.1E-11	5.6E-11

a Dose coefficients for this element are based on age-specific biokinetic data

b The f_1 value for 1 to 15 year olds is 0.4

Table A.1.—(continued)

Nuclide	Physical half-life	f_1 <1y	e(τ) 3 months	f_1 ≥1y	e(τ) 1 Year	5 Years	10 Years	15 Years	Adult
Y-87	3.35 d	0.001	4.6E-09	1.0E-04	3.2E-09	1.8E-09	1.1E-09	7.0E-10	5.5E-10
Y-88	107 d	0.001	8.1E-09	1.0E-04	6.0E-09	3.5E-09	2.4E-09	1.6E-09	1.3E-09
Y-90	2.67 d	0.001	3.1E-08	1.0E-04	2.0E-08	1.0E-08	5.9E-09	3.3E-09	2.7E-09
Y-90m	3.19 h	0.001	1.8E-09	1.0E-04	1.2E-09	6.1E-10	3.7E-10	2.2E-10	1.7E-10
Y-91	58.5 d	0.001	2.8E-08	1.0E-04	1.8E-08	8.8E-09	5.2E-09	2.9E-09	2.4E-09
Y-91m	0.828 h	0.001	9.2E-11	1.0E-04	6.0E-11	3.3E-11	2.1E-11	1.4E-11	1.1E-11
Y-92	3.54 h	0.001	5.9E-09	1.0E-04	3.6E-09	1.8E-09	1.0E-09	6.2E-10	4.9E-10
Y-93	10.1 h	0.001	1.4E-08	1.0E-04	8.5E-09	4.3E-09	2.5E-09	1.4E-09	1.2E-09
Y-94	0.318 h	0.001	9.9E-10	1.0E-04	5.5E-10	2.7E-10	1.5E-10	1.0E-10	8.1E-11
Y-95	0.178 h	0.001	5.7E-10	1.0E-04	3.1E-10	1.5E-10	8.7E-11	5.9E-11	4.6E-11
Zirconium*									
Zr-86	16.5 h	0.020	6.9E-09	0.010	4.8E-09	2.7E-09	1.7E-09	1.1E-09	8.6E-10
Zr-88	83.4 d	0.020	2.8E-09	0.010	2.0E-09	1.2E-09	8.0E-10	5.4E-10	4.5E-10
Zr-89	3.27 d	0.020	6.5E-09	0.010	4.5E-09	2.5E-09	1.6E-09	9.9E-10	7.9E-10
Zr-93	1.53E+06 y	0.020	1.2E-09	0.010	7.6E-10	5.1E-10	5.8E-10	9.6E-10	1.1E-09
Zr-95	64.0 d	0.020	8.5E-09	0.010	5.6E-09	3.0E-09	1.9E-09	1.2E-09	9.5E-10
Zr-97	16.9 h	0.020	2.2E-08	0.010	1.4E-08	7.3E-09	4.4E-09	2.6E-09	2.1E-09
Niobium*									
Nb-88	0.238 h	0.020	6.7E-10	0.010	3.8E-10	1.9E-10	1.1E-10	7.9E-11	6.5E-11
Nb-89	2.03 h	0.020	3.0E-09	0.010	2.0E-09	1.0E-09	6.0E-10	3.4E-10	2.7E-10
Nb-89	1.10 h	0.020	1.5E-09	0.010	8.7E-10	4.4E-10	2.7E-10	1.8E-10	1.4E-10
Nb-90	14.6 h	0.020	1.1E-08	0.010	7.2E-09	3.9E-09	2.5E-09	1.6E-09	1.3E-09
Nb-93m	13.6 y	0.020	1.5E-09	0.010	9.1E-10	4.6E-10	2.7E-10	1.5E-10	1.2E-10
Nb-94	2.03E+04 y	0.020	1.5E-08	0.010	9.7E-09	5.3E-09	3.4E-09	2.1E-09	1.7E-09
Nb-95	35.1 d	0.020	4.6E-09	0.010	3.2E-09	1.8E-09	1.1E-09	7.4E-10	5.3E-10
Nb-95m	3.61 d	0.020	6.4E-09	0.010	4.1E-09	2.1E-09	1.2E-09	7.1E-10	5.6E-10
Nb-96	23.3 h	0.020	9.2E-09	0.010	6.3E-09	3.4E-09	2.2E-09	1.4E-09	1.1E-09
Nb-97	1.20 h	0.020	7.7E-10	0.010	4.5E-10	2.3E-10	1.3E-10	8.7E-11	6.8E-11
Nb-98	0.858 h	0.020	1.2E-09	0.010	7.1E-10	3.6E-10	2.2E-10	1.4E-10	1.1E-10

* Dose coefficients for this element are based on age-specific biokinetic data

Table A.1.—(continued)

Nuclide	Physical half-life	f_1 <1y	e(τ) 3 months	f_1 ≥1y	e(τ) 1 Year	5 Years	10 Years	15 Years	Adult
Molybdenum[a]									
Mo-90	5.67 h	1.000	1.7E-09	1.000	1.2E-09	6.3E-10	4.0E-10	2.7E-10	2.2E-10
Mo-93	3.50E+03 y	1.000	7.9E-09	1.000	6.9E-09	5.0E-09	4.0E-09	3.4E-09	3.1E-09
Mo-93m	6.85 h	1.000	8.0E-10	1.000	5.4E-10	3.1E-10	2.0E-10	1.4E-10	1.1E-10
Mo-99	2.75 d	1.000	5.5E-09	1.000	3.5E-09	1.8E-09	1.1E-09	7.6E-10	6.0E-10
Mo-101	0.244 h	1.000	4.8E-10	1.000	2.7E-10	1.3E-10	7.6E-11	5.2E-11	4.1E-11
Technetium[a]									
Tc-93	2.75 h	1.000	2.7E-10	0.500	2.5E-10	1.5E-10	9.8E-11	6.8E-11	5.5E-11
Tc-93m	0.725 h	1.000	2.0E-10	0.500	1.3E-10	7.3E-11	4.6E-11	3.2E-11	2.5E-11
Tc-94	4.88 h	1.000	1.2E-09	0.500	1.0E-09	5.8E-10	3.7E-10	2.5E-10	2.0E-10
Tc-94m	0.867 h	1.000	1.3E-09	0.500	6.5E-10	3.3E-10	1.9E-10	1.3E-10	1.0E-10
Tc-95	20.0 h	1.000	9.9E-10	0.500	8.7E-10	5.0E-10	3.3E-10	2.3E-10	1.8E-10
Tc-95m	61.0 d	1.000	4.7E-09	0.500	2.8E-09	1.6E-09	1.0E-09	7.0E-10	5.6E-10
Tc-96	4.28 d	1.000	6.7E-09	0.500	5.1E-09	3.0E-09	2.0E-09	1.4E-09	1.1E-09
Tc-96m	0.858 h	1.000	1.0E-10	0.500	6.5E-11	3.6E-11	2.3E-11	1.6E-11	1.2E-11
Tc-97	2.60E+06 y	1.000	9.9E-10	0.500	4.9E-10	2.4E-10	1.4E-10	8.8E-11	6.8E-11
Tc-97m	87.0 d	1.000	8.7E-09	0.500	4.1E-09	2.0E-09	1.1E-09	7.0E-10	5.5E-10
Tc-98	4.20E+06 y	1.000	2.3E-08	0.500	1.2E-08	6.1E-09	3.7E-09	2.5E-09	2.0E-09
Tc-99	2.13E+05 y	1.000	1.0E-08	0.500	4.8E-09	2.3E-09	1.3E-09	8.2E-10	6.4E-10
Tc-99m	6.02 h	1.000	2.0E-10	0.500	1.3E-10	7.2E-11	4.3E-11	2.8E-11	2.2E-11
Tc-101	0.237 h	1.000	2.4E-10	0.500	1.3E-10	6.1E-11	3.5E-11	2.4E-11	1.9E-11
Tc-104	0.303 h	1.000	1.0E-09	0.500	5.3E-10	2.6E-10	1.5E-10	1.0E-10	8.0E-11
Ruthenium[a]									
Ru-94	0.863 h	0.100	9.3E-10	0.050	5.9E-10	3.1E-10	1.9E-10	1.2E-10	9.4E-11
Ru-97	2.90 d	0.100	1.2E-09	0.050	8.5E-10	4.7E-10	3.0E-10	1.9E-10	1.5E-10
Ru-103	39.3 d	0.100	7.1E-09	0.050	4.6E-09	2.4E-09	1.5E-09	9.2E-10	7.3E-10
Ru-105	4.44 h	0.100	2.7E-09	0.050	1.8E-09	9.1E-10	5.5E-10	3.3E-10	2.6E-10
Ru-106	1.01 y	0.100	8.4E-08	0.050	4.9E-08	2.5E-08	1.5E-08	8.6E-09	7.0E-09

a Dose coefficients for this element are based on age-specific biokinetic data

Table A.1.—(continued)

Nuclide	Physical half-life	f_1 <1y	$e(\tau)$ 3 months	f_1 ≥1y	$e(\tau)$ 1 Year	5 Years	10 Years	15 Years	Adult
Rhodium									
Rh-99	16.0 d	0.100	4.2E-09	0.050	2.9E-09	1.6E-09	1.0E-09	6.5E-10	5.1E-10
Rh-99m	4.70 h	0.100	4.9E-10	0.050	3.5E-10	2.0E-10	1.3E-10	8.3E-11	6.6E-11
Rh-100	20.8 h	0.100	4.9E-09	0.050	3.6E-09	2.0E-09	1.4E-09	8.8E-10	7.1E-10
Rh-101	3.20 y	0.100	4.9E-09	0.050	2.8E-09	1.6E-09	1.0E-09	6.7E-10	5.5E-10
Rh-101m	4.34 d	0.100	1.7E-09	0.050	1.2E-09	6.8E-10	4.4E-10	2.8E-10	2.2E-10
Rh-102	2.90 y	0.100	1.9E-08	0.050	1.0E-08	6.4E-09	4.3E-09	3.0E-09	2.6E-09
Rh-102m	207 d	0.100	1.2E-08	0.050	7.4E-09	3.9E-09	2.4E-09	1.4E-09	1.2E-09
Rh-103m	0.935 h	0.100	4.7E-11	0.050	2.7E-11	1.3E-11	7.4E-12	4.8E-12	3.8E-12
Rh-105	1.47 d	0.100	4.0E-09	0.050	2.7E-09	1.3E-09	8.0E-10	4.6E-10	3.7E-10
Rh-106m	2.20 h	0.100	1.4E-09	0.050	9.7E-10	5.3E-10	3.3E-10	2.0E-10	1.6E-10
Rh-107	0.362 h	0.100	2.9E-10	0.050	1.6E-10	7.9E-11	4.5E-11	3.1E-11	2.4E-11
Palladium									
Pd-100	3.63 d	0.050	7.4E-09	0.005	5.2E-09	2.9E-09	1.9E-09	1.2E-09	9.4E-10
Pd-101	8.27 h	0.050	8.2E-10	0.005	5.7E-10	3.1E-10	1.9E-10	1.2E-10	9.4E-11
Pd-103	17.0 d	0.050	2.2E-09	0.005	1.4E-09	7.2E-10	4.3E-10	2.4E-10	1.9E-10
Pd-107	6.50E+06 y	0.050	4.4E-10	0.005	2.8E-10	1.4E-10	8.1E-11	4.6E-11	3.7E-11
Pd-109	13.4 h	0.050	6.3E-09	0.005	4.1E-09	2.0E-09	1.2E-09	6.8E-10	5.5E-10
Silver[a]									
Ag-102	0.215 h	0.100	4.2E-10	0.050	2.4E-10	1.2E-10	7.3E-11	5.0E-11	4.0E-11
Ag-103	1.09 h	0.100	4.5E-10	0.050	2.7E-10	1.4E-10	8.3E-11	5.5E-11	4.3E-11
Ag-104	1.15 h	0.100	4.3E-10	0.050	2.9E-10	1.7E-10	1.1E-10	7.5E-11	6.0E-11
Ag-104m	0.558 h	0.100	5.6E-10	0.050	3.3E-10	1.7E-10	1.0E-10	6.8E-11	5.4E-11
Ag-105	41.0 d	0.100	3.9E-09	0.050	2.5E-09	1.4E-09	9.1E-10	5.9E-10	4.7E-10
Ag-106	0.399 h	0.100	3.7E-10	0.050	2.1E-10	1.0E-10	6.0E-11	4.1E-11	3.2E-11
Ag-106m	8.41 d	0.100	9.7E-09	0.050	6.9E-09	4.1E-09	2.8E-09	1.8E-09	1.5E-09
Ag-108m	1.27E+02 y	0.100	2.1E-08	0.050	1.1E-08	6.5E-09	4.3E-09	2.8E-09	2.3E-09
Ag-110m	250 d	0.100	2.4E-08	0.050	1.4E-08	7.8E-09	5.2E-09	3.4E-09	2.8E-09
Ag-111	7.45 d	0.100	1.4E-08	0.050	9.3E-09	4.6E-09	2.7E-09	1.6E-09	1.3E-09
Ag-112	3.12 h	0.100	4.9E-09	0.050	3.0E-09	1.5E-09	8.9E-10	5.4E-10	4.3E-10

a Dose coefficients for this element are based on age-specific biokinetic data

Table A.1.—(*continued*)

Nuclide	Physical half-life	f_1 <1y	$e(\tau)$ 3 months	f_1 ≥1y	$e(\tau)$ 1 Year	5 Years	10 Years	15 Years	Adult
Ag-115	0.333 h	0.100	7.2E-10	0.050	4.1E-10	2.0E-10	1.2E-10	7.7E-11	6.0E-11
Cadmium									
Cd-104	0.961 h	0.100	4.2E-10	0.050	2.9E-10	1.7E-10	1.1E-10	7.2E-11	5.4E-11
Cd-107	6.49 h	0.100	7.1E-10	0.050	4.6E-10	2.3E-10	1.3E-10	7.8E-11	6.2E-11
Cd-109	1.27 y	0.100	2.1E-08	0.050	9.5E-09	5.5E-09	3.5E-09	2.4E-09	2.0E-09
Cd-113	9.30E+15 y	0.100	1.0E-07	0.050	4.8E-08	3.7E-08	3.0E-08	2.6E-08	2.5E-08
Cd-113m	13.6 y	0.100	1.2E-07	0.050	5.6E-08	3.9E-08	2.9E-08	2.4E-08	2.3E-08
Cd-115	2.23 d	0.100	1.4E-08	0.050	9.7E-09	4.9E-09	2.9E-09	1.7E-09	1.4E-09
Cd-115m	44.6 d	0.100	4.1E-08	0.050	1.9E-08	9.7E-09	6.9E-09	4.1E-09	3.3E-09
Cd-117	2.49 h	0.100	2.9E-09	0.050	1.9E-09	9.5E-10	5.7E-10	3.5E-10	2.8E-10
Cd-117m	3.36 h	0.100	2.6E-09	0.050	1.7E-09	9.0E-10	5.6E-10	3.5E-10	2.8E-10
Indium									
In-109	4.20 h	0.040	5.2E-10	0.020	3.6E-10	2.0E-10	1.3E-10	8.2E-11	6.6E-11
In-110	4.90 h	0.040	1.5E-09	0.020	1.1E-09	6.5E-10	4.4E-10	3.0E-10	2.4E-10
In-110	1.15 h	0.040	1.1E-09	0.020	6.4E-10	3.2E-10	1.9E-10	1.3E-10	1.0E-10
In-111	2.83 d	0.040	2.4E-09	0.020	1.7E-09	9.1E-10	5.9E-10	3.7E-10	2.9E-10
In-112	0.240 h	0.040	1.2E-10	0.020	6.7E-11	3.3E-11	1.9E-11	1.3E-11	1.0E-11
In-113m	1.66 h	0.040	3.0E-10	0.020	1.8E-10	9.3E-11	6.2E-11	3.6E-11	2.8E-11
In-114m	49.5 d	0.040	5.6E-08	0.020	3.1E-08	1.5E-08	9.0E-09	5.2E-09	4.1E-09
In-115	5.10E+15 y	0.040	1.3E-07	0.020	6.4E-08	4.8E-08	4.3E-08	3.6E-08	3.2E-08
In-115m	4.49 h	0.040	9.6E-10	0.020	6.0E-10	3.0E-10	1.8E-10	1.1E-10	8.6E-11
In-116m	0.902 h	0.040	5.8E-10	0.020	3.6E-10	1.9E-10	1.2E-10	8.0E-11	6.4E-11
In-117	0.730 h	0.040	3.3E-10	0.020	1.9E-10	9.7E-11	5.8E-11	3.9E-11	3.1E-11
In-117m	1.94 h	0.040	1.4E-09	0.020	8.6E-10	4.3E-10	2.5E-10	1.6E-10	1.2E-10
In-119m	0.300 h	0.040	5.9E-10	0.020	3.2E-10	1.6E-10	8.8E-11	6.0E-11	4.7E-11
Tin									
Sn-110	4.00 h	0.040	3.5E-09	0.020	2.3E-09	1.2E-09	7.4E-10	4.4E-10	3.5E-10
Sn-111	0.588 h	0.040	2.5E-10	0.020	1.5E-10	7.4E-11	4.4E-11	3.0E-11	2.3E-11
Sn-113	115 d	0.040	7.8E-09	0.020	5.0E-09	2.6E-09	1.6E-09	9.2E-10	7.3E-10
Sn-117m	13.6 d	0.040	7.7E-09	0.020	5.0E-09	2.5E-09	1.5E-09	8.8E-10	7.1E-10

Table A.1.—(continued)

Nuclide	Physical half-life	f_1 <1y	e(τ) 3 months	f_1 ≥1y	e(τ) 1 Year	5 Years	10 Years	15 Years	Adult
Sn-119m	293 d	0.040	4.1E-09	0.020	2.5E-09	1.3E-09	7.5E-10	4.3E-10	3.4E-10
Sn-121	1.13 d	0.040	2.6E-09	0.020	1.7E-09	8.4E-10	5.0E-10	2.8E-10	2.3E-10
Sn-121m	55.0 y	0.040	4.6E-09	0.020	2.7E-09	1.4E-09	8.2E-10	4.7E-10	3.8E-10
Sn-123	129 d	0.040	2.5E-08	0.020	1.6E-08	7.8E-09	4.6E-09	2.6E-09	2.1E-09
Sn-123m	0.668 h	0.040	4.7E-10	0.020	2.6E-10	1.3E-10	7.3E-11	4.9E-11	3.8E-11
Sn-125	9.64 d	0.040	3.5E-08	0.020	2.2E-08	1.1E-08	6.7E-09	3.8E-09	3.1E-09
Sn-126	1.00E+05 y	0.040	5.0E-08	0.020	3.0E-08	1.6E-08	9.8E-09	5.9E-09	4.7E-09
Sn-127	2.10 h	0.040	2.0E-09	0.020	1.3E-09	6.6E-10	4.0E-10	2.5E-10	2.0E-10
Sn-128	0.985 h	0.040	1.6E-09	0.020	9.7E-10	4.9E-10	3.0E-10	1.9E-10	1.5E-10
Antimony[a]									
Sb-115	0.530 h	0.200	2.5E-10	0.100	1.5E-10	7.5E-11	4.5E-11	3.1E-11	2.4E-11
Sb-116	0.263 h	0.200	2.7E-10	0.100	1.6E-10	8.0E-11	4.8E-11	3.3E-11	2.6E-11
Sb-116m	1.00 h	0.200	5.0E-10	0.100	3.3E-10	1.9E-10	1.2E-10	8.3E-11	6.7E-11
Sb-117	2.80 h	0.200	1.6E-10	0.100	1.0E-10	5.6E-11	3.5E-11	2.2E-11	1.8E-11
Sb-118m	5.00 h	0.200	1.3E-09	0.100	1.0E-09	5.8E-10	3.9E-10	2.6E-10	2.1E-10
Sb-119	1.59 d	0.200	8.4E-10	0.100	5.8E-10	3.0E-10	1.8E-10	1.0E-10	8.0E-11
Sb-120	5.76 d	0.200	8.1E-09	0.100	6.0E-09	3.5E-09	2.3E-09	1.6E-09	1.2E-09
Sb-120	0.265 h	0.200	1.7E-10	0.100	9.4E-11	4.6E-11	2.7E-11	1.8E-11	1.4E-11
Sb-122	2.70 d	0.200	1.8E-08	0.100	1.2E-08	6.1E-09	3.7E-09	2.1E-09	1.7E-09
Sb-124	60.2 d	0.200	2.5E-08	0.100	1.6E-08	8.4E-09	5.2E-09	3.2E-09	2.5E-09
Sb-124m	0.337 h	0.200	8.5E-11	0.100	4.9E-11	2.5E-11	1.5E-11	1.0E-11	8.0E-12
Sb-125	2.77 y	0.200	1.1E-08	0.100	6.1E-09	3.4E-09	2.1E-09	1.4E-09	1.1E-09
Sb-126	12.4 d	0.200	2.0E-08	0.100	1.4E-08	7.6E-09	4.9E-09	3.1E-09	2.4E-09
Sb-126m	0.317 h	0.200	3.9E-10	0.100	2.2E-10	1.1E-10	6.6E-11	4.5E-11	3.6E-11
Sb-127	3.85 d	0.200	1.7E-08	0.100	1.2E-08	5.9E-09	3.6E-09	2.1E-09	1.7E-09
Sb-128	9.01 h	0.200	6.3E-09	0.100	4.5E-09	2.4E-09	1.5E-09	9.5E-10	7.6E-10
Sb-128	0.173 h	0.200	3.7E-10	0.100	2.1E-10	1.0E-10	6.0E-11	4.1E-11	3.3E-11
Sb-129	4.32 h	0.200	4.3E-09	0.100	2.8E-09	1.5E-09	8.8E-10	5.3E-10	4.2E-10
Sb-130	0.667 h	0.200	9.1E-10	0.100	5.4E-10	2.8E-10	1.7E-10	1.2E-10	9.1E-11
Sb-131	0.383 h	0.200	1.1E-09	0.100	7.3E-10	3.9E-10	2.1E-10	1.4E-10	1.1E-10

a Dose coefficients for this element are based on age-specific biokinetic data

Table A.1.—(continued)

Nuclide	Physical half-life	f_1 <1y	$e(\tau)$ 3 months	f_1 ≥1y	$e(\tau)$ 1 Year	5 Years	10 Years	15 Years	Adult
Tellurium[a]									
Te-116	2.49 h	0.600	1.4E-09	0.300	1.0E-09	5.5E-10	3.4E-10	2.1E-10	1.7E-10
Te-121	17.0 d	0.600	3.1E-09	0.300	2.0E-09	1.2E-09	8.0E-10	5.4E-10	4.3E-10
Te-121m	154 d	0.600	2.7E-08	0.300	1.2E-08	6.9E-09	4.2E-09	2.8E-09	2.3E-09
Te-123	1.00E+13 y	0.600	2.0E-08	0.300	9.3E-09	6.9E-09	5.4E-09	4.7E-09	4.4E-09
Te-123m	120 d	0.600	1.9E-08	0.300	8.8E-09	4.9E-09	2.8E-09	1.7E-09	1.4E-09
Te-125m	58.0 d	0.600	1.3E-08	0.300	6.3E-09	3.3E-09	1.9E-09	1.1E-09	8.7E-10
Te-127	9.35 h	0.600	1.5E-09	0.300	1.2E-09	6.2E-10	3.6E-10	2.1E-10	1.7E-10
Te-127m	109 d	0.600	4.1E-08	0.300	1.8E-08	9.5E-09	5.2E-09	3.0E-09	2.3E-09
Te-129	1.16 h	0.600	7.5E-10	0.300	4.4E-10	2.1E-10	1.2E-10	8.0E-11	6.3E-11
Te-129m	33.6 d	0.600	4.4E-08	0.300	2.4E-08	1.2E-08	6.6E-09	3.9E-09	3.0E-09
Te-131	0.417 h	0.600	9.0E-10	0.300	6.6E-10	3.5E-10	1.9E-10	1.2E-10	8.7E-11
Te-131m	1.25 d	0.600	2.0E-08	0.300	1.4E-08	7.8E-09	4.3E-09	2.7E-09	1.9E-09
Te-132	3.26 d	0.600	4.8E-08	0.300	3.0E-08	1.6E-08	8.3E-09	5.3E-09	3.8E-09
Te-133	0.207 h	0.600	8.4E-10	0.300	6.3E-10	3.3E-10	1.6E-10	1.1E-10	7.2E-11
Te-133m	0.923 h	0.600	3.1E-09	0.300	2.4E-09	1.3E-09	6.3E-10	4.1E-10	2.8E-10
Te-134	0.696 h	0.600	1.1E-09	0.300	7.5E-10	3.9E-10	2.2E-10	1.4E-10	1.1E-10
Iodine[a]									
I-120	1.35 h	1.000	3.9E-09	1.000	2.8E-09	1.4E-09	7.2E-10	4.8E-10	3.4E-10
I-120m	0.863 h	1.000	2.3E-09	1.000	1.5E-09	7.8E-10	4.2E-10	2.9E-10	2.1E-10
I-121	2.12 h	1.000	6.2E-10	1.000	5.3E-10	3.1E-10	1.7E-10	1.2E-10	8.2E-11
I-123	13.2 h	1.000	2.2E-09	1.000	1.9E-09	1.1E-09	4.9E-10	3.3E-10	2.1E-10
I-124	4.18 d	1.000	1.2E-07	1.000	1.1E-07	6.3E-08	3.1E-08	2.0E-08	1.3E-08
I-125	60.1 d	1.000	5.2E-08	1.000	5.7E-08	4.1E-08	3.1E-08	2.2E-08	1.5E-08
I-126	13.0 d	1.000	2.1E-07	1.000	2.1E-07	1.3E-07	6.8E-08	4.5E-08	2.9E-08
I-128	0.416 h	1.000	5.7E-10	1.000	3.3E-10	1.6E-10	8.9E-11	6.0E-11	4.6E-11
I-129	1.57E+07 y	1.000	1.8E-07	1.000	2.2E-07	1.7E-07	1.9E-07	1.4E-07	1.1E-07
I-130	12.4 h	1.000	2.1E-08	1.000	1.8E-08	9.8E-09	4.6E-09	3.0E-09	2.0E-09
I-131	8.04 d	1.000	1.8E-07	1.000	1.8E-07	1.0E-07	5.2E-08	3.4E-08	2.2E-08
I-132	2.30 h	1.000	3.0E-09	1.000	2.4E-09	1.3E-09	6.2E-10	4.1E-10	2.9E-10

a Dose coefficients for this element are based on age-specific biokinetic data

Table A.1.—(continued)

Nuclide	Physical half-life	f_1 <1y	$e(\tau)$ 3 months	f_1 ≥1y	$e(\tau)$ 1 Year	5 Years	10 Years	15 Years	Adult
I-132m	1.39 h	1.000	2.4E-09	1.000	2.0E-09	1.1E-09	5.0E-10	3.3E-10	2.2E-10
I-133	20.8 h	1.000	4.9E-08	1.000	4.4E-08	2.3E-08	1.0E-08	6.8E-09	4.3E-09
I-134	0.876 h	1.000	1.1E-09	1.000	7.5E-10	3.9E-10	2.1E-10	1.4E-10	1.1E-10
I-135	6.61 h	1.000	1.0E-08	1.000	8.9E-09	4.7E-09	2.2E-09	1.4E-09	9.3E-10
Caesium[a]									
Cs-125	0.750 h	1.000	3.9E-10	1.000	2.2E-10	1.1E-10	6.5E-11	4.4E-11	3.5E-11
Cs-127	6.25 h	1.000	1.8E-10	1.000	1.2E-10	6.6E-11	4.2E-11	2.9E-11	2.4E-11
Cs-129	1.34 d	1.000	4.4E-10	1.000	3.0E-10	1.7E-10	1.1E-10	7.2E-11	6.0E-11
Cs-130	0.498 h	1.000	3.3E-10	1.000	1.8E-10	9.0E-11	5.2E-11	3.6E-11	2.8E-11
Cs-131	9.69 d	1.000	4.6E-10	1.000	2.9E-10	1.6E-10	1.0E-10	6.9E-11	5.8E-11
Cs-132	6.48 d	1.000	2.7E-09	1.000	1.8E-09	1.1E-09	7.7E-10	5.7E-10	5.0E-10
Cs-134	2.06 y	1.000	2.6E-08	1.000	1.6E-08	1.3E-08	1.4E-08	1.9E-08	1.9E-08
Cs-134m	2.90 h	1.000	2.1E-10	1.000	1.2E-10	5.9E-11	3.5E-11	2.5E-11	2.0E-11
Cs-135	2.30E+06 y	1.000	4.1E-09	1.000	2.3E-09	1.7E-09	1.7E-09	2.0E-09	2.0E-09
Cs-135m	0.883 h	1.000	1.3E-10	1.000	8.6E-11	4.9E-11	3.2E-11	2.3E-11	1.9E-11
Cs-136	13.1 d	1.000	1.5E-08	1.000	9.5E-09	6.1E-09	4.4E-09	3.4E-09	3.0E-09
Cs-137	30.0 y	1.000	2.1E-08	1.000	1.2E-08	9.6E-09	1.0E-08	1.3E-08	1.3E-08
Cs-138	0.536 h	1.000	1.1E-09	1.000	5.9E-10	2.9E-10	1.7E-10	1.2E-10	9.2E-11
Barium[a,b]									
Ba-126	1.61 h	0.600	2.7E-09	0.200	1.7E-09	8.5E-10	5.0E-10	3.1E-10	2.6E-10
Ba-128	2.43 d	0.600	2.0E-08	0.200	1.7E-08	9.0E-09	5.2E-09	3.0E-09	2.7E-09
Ba-131	11.8 d	0.600	4.2E-09	0.200	2.6E-09	1.4E-09	9.4E-10	6.2E-10	4.5E-10
Ba-131m	0.243 h	0.600	5.8E-11	0.200	3.2E-11	1.6E-11	9.3E-12	6.3E-12	4.9E-12
Ba-133	10.7 y	0.600	2.2E-08	0.200	6.2E-09	3.9E-09	4.6E-09	7.3E-09	1.5E-09
Ba-133m	1.62 d	0.600	4.2E-09	0.200	3.6E-09	1.8E-09	1.1E-09	5.9E-10	5.4E-10
Ba-135m	1.20 d	0.600	3.3E-09	0.200	2.9E-09	1.5E-09	8.5E-10	4.7E-10	4.3E-10
Ba-139	1.38 h	0.600	1.4E-09	0.200	8.4E-10	4.1E-10	2.4E-10	1.5E-10	1.2E-10
Ba-140	12.7 d	0.600	3.2E-08	0.200	1.8E-08	9.2E-09	5.8E-09	3.7E-09	2.6E-09
Ba-141	0.305 h	0.600	7.6E-10	0.200	4.7E-10	2.3E-10	1.3E-10	8.6E-11	7.0E-11
Ba-142	0.177 h	0.600	3.6E-10	0.200	2.2E-10	1.1E-10	6.6E-11	4.3E-11	3.5E-11

a Dose coefficients for this element are based on age-specific biokinetic data

b The f_1 value for 1 to 15 year olds is 0.3

Table A.1.—(*continued*)

Nuclide	Physical half-life	f_1 <1y	$e(\tau)$ 3 months	f_1 ≥1y	$e(\tau)$ 1 Year	5 Years	10 Years	15 Years	Adult
Lanthanum									
La-131	0.983 h	0.005	3.5E-10	5.0E-04	2.1E-10	1.1E-10	6.6E-11	4.4E-11	3.5E-11
La-132	4.80 h	0.005	3.8E-09	5.0E-04	2.4E-09	1.3E-09	7.8E-10	4.8E-10	3.9E-10
La-135	19.5 h	0.005	2.8E-10	5.0E-04	1.9E-10	1.0E-10	6.4E-11	3.9E-11	3.0E-11
La-137	6.00E+04 y	0.005	1.1E-09	5.0E-04	4.5E-10	2.5E-10	1.6E-10	1.0E-10	8.1E-11
La-138	1.35E+11 y	0.005	1.3E-08	5.0E-04	4.6E-09	2.7E-09	1.9E-09	1.3E-09	1.1E-09
La-140	1.68 d	0.005	2.0E-08	5.0E-04	1.3E-08	6.8E-09	4.2E-09	2.5E-09	2.0E-09
La-141	3.93 h	0.005	4.3E-09	5.0E-04	2.6E-09	1.3E-09	7.6E-10	4.5E-10	3.6E-10
La-142	1.54 h	0.005	1.9E-09	5.0E-04	1.1E-09	5.8E-10	3.5E-10	2.3E-10	1.8E-10
La-143	0.237 h	0.005	6.9E-10	5.0E-04	3.9E-10	1.9E-10	1.1E-10	7.1E-11	5.6E-11
Cerium[a]									
Ce-134	3.00 d	0.005	2.8E-08	5.0E-04	1.8E-08	9.1E-09	5.5E-09	3.2E-09	2.5E-09
Ce-135	17.6 h	0.005	7.0E-09	5.0E-04	4.7E-09	2.6E-09	1.6E-09	1.0E-09	7.9E-10
Ce-137	9.00 h	0.005	2.6E-10	5.0E-04	1.7E-10	8.8E-11	5.4E-11	3.2E-11	2.5E-11
Ce-137m	1.43 d	0.005	6.1E-09	5.0E-04	3.9E-09	2.0E-09	1.2E-09	6.8E-10	5.4E-10
Ce-139	138 d	0.005	2.6E-09	5.0E-04	1.6E-09	8.6E-10	5.4E-10	3.3E-10	2.6E-10
Ce-141	32.5 d	0.005	8.1E-09	5.0E-04	5.1E-09	2.6E-09	1.5E-09	8.8E-10	7.1E-10
Ce-143	1.38 d	0.005	1.2E-08	5.0E-04	8.0E-09	4.1E-09	2.4E-09	1.4E-09	1.1E-09
Ce-144	284 d	0.005	6.6E-08	5.0E-04	3.9E-08	1.9E-08	1.1E-08	6.5E-09	5.2E-09
Praseodymium									
Pr-136	0.218 h	0.005	3.7E-10	5.0E-04	2.1E-10	1.0E-10	6.1E-11	4.2E-11	3.3E-11
Pr-137	1.28 h	0.005	4.1E-10	5.0E-04	2.5E-10	1.3E-10	7.7E-11	5.0E-11	4.0E-11
Pr-138m	2.10 h	0.005	1.0E-09	5.0E-04	7.4E-10	4.1E-10	2.6E-10	1.6E-10	1.3E-10
Pr-139	4.51 h	0.005	3.2E-10	5.0E-04	2.0E-10	1.1E-10	6.5E-11	4.0E-11	3.1E-11
Pr-142	19.1 h	0.005	1.5E-08	5.0E-04	9.8E-09	4.9E-09	2.9E-09	1.6E-09	1.3E-09
Pr-142m	0.243 h	0.005	2.0E-10	5.0E-04	1.2E-10	6.2E-11	3.7E-11	2.1E-11	1.7E-11
Pr-143	13.6 d	0.005	1.4E-08	5.0E-04	8.7E-09	4.3E-09	2.6E-09	1.5E-09	1.2E-09
Pr-144	0.288 h	0.005	6.4E-10	5.0E-04	3.5E-10	1.7E-10	9.5E-11	6.5E-11	5.0E-11
Pr-145	5.98 h	0.005	4.7E-09	5.0E-04	2.9E-09	1.4E-09	8.5E-10	4.9E-10	3.9E-10
Pr-147	0.227 h	0.005	3.9E-10	5.0E-04	2.2E-10	1.1E-10	6.1E-11	4.2E-11	3.3E-11

a   Dose coefficients for this element are based on age-specific biokinetic data

Table A.1.—(*continued*)

Nuclide	Physical half-life	f_1 <1y	e(τ) 3 months	f_1 ≥1y	e(τ) 1 Year	5 Years	10 Years	15 Years	Adult
Neodymium									
Nd-136	0.844 h	0.005	1.0E-09	5.0E-04	6.1E-10	3.1E-10	1.9E-10	1.2E-10	9.9E-11
Nd-138	5.04 h	0.005	7.2E-09	5.0E-04	4.5E-09	2.3E-09	1.3E-09	8.0E-10	6.4E-10
Nd-139	0.495 h	0.005	2.1E-10	5.0E-04	1.2E-10	6.3E-11	3.7E-11	2.5E-11	2.0E-11
Nd-139m	5.50 h	0.005	2.1E-09	5.0E-04	1.4E-09	7.8E-10	5.0E-10	3.1E-10	2.5E-10
Nd-141	2.49 h	0.005	7.8E-11	5.0E-04	5.0E-11	2.7E-11	1.6E-11	1.0E-11	8.3E-12
Nd-147	11.0 d	0.005	1.2E-08	5.0E-04	7.8E-09	3.9E-09	2.3E-09	1.3E-09	1.1E-09
Nd-149	1.73 h	0.005	1.4E-09	5.0E-04	8.7E-10	4.3E-10	2.6E-10	1.6E-10	1.2E-10
Nd-151	0.207 h	0.005	3.4E-10	5.0E-04	2.0E-10	9.7E-11	5.7E-11	3.8E-11	3.0E-11
Promethium									
Pm-141	0.348 h	0.005	4.2E-10	5.0E-04	2.4E-10	1.2E-10	6.8E-11	4.6E-11	3.6E-11
Pm-143	265 d	0.005	1.9E-09	5.0E-04	1.2E-09	6.7E-10	4.4E-10	2.9E-10	2.3E-10
Pm-144	363 d	0.005	7.6E-09	5.0E-04	4.7E-09	2.7E-09	1.8E-09	1.2E-09	9.7E-10
Pm-145	17.7 y	0.005	1.5E-09	5.0E-04	6.8E-10	3.7E-10	2.3E-10	1.4E-10	1.1E-10
Pm-146	5.53 y	0.005	1.0E-08	5.0E-04	5.1E-09	2.8E-09	1.8E-09	1.1E-09	9.0E-10
Pm-147	2.62 y	0.005	3.6E-09	5.0E-04	1.9E-09	9.6E-10	5.7E-10	3.2E-10	2.6E-10
Pm-148	5.37 d	0.005	3.0E-08	5.0E-04	1.9E-08	9.7E-09	5.8E-09	3.3E-09	2.7E-09
Pm-148m	41.3 d	0.005	1.5E-08	5.0E-04	1.0E-08	5.5E-09	3.5E-09	2.2E-09	1.7E-09
Pm-149	2.21 d	0.005	1.2E-08	5.0E-04	7.4E-09	3.7E-09	2.2E-09	1.2E-09	9.9E-10
Pm-150	2.68 h	0.005	2.8E-09	5.0E-04	1.7E-09	8.7E-10	5.2E-10	3.2E-10	2.6E-10
Pm-151	1.18 d	0.005	8.0E-09	5.0E-04	5.1E-09	2.6E-09	1.6E-09	9.1E-10	7.3E-10
Samarium									
Sm-141	0.170 h	0.005	4.5E-10	5.0E-04	2.5E-10	1.3E-10	7.3E-11	5.0E-11	3.9E-11
Sm-141m	0.377 h	0.005	7.0E-10	5.0E-04	4.0E-10	2.0E-10	1.2E-10	8.2E-11	6.5E-11
Sm-142	1.21 h	0.005	2.2E-09	5.0E-04	1.3E-09	6.2E-10	3.6E-10	2.4E-10	1.9E-10
Sm-145	340 d	0.005	2.4E-09	5.0E-04	1.4E-09	7.3E-10	4.5E-10	2.7E-10	2.1E-10
Sm-146	1.03E+08 y	0.005	1.5E-06	5.0E-04	1.5E-07	1.0E-07	7.0E-08	5.8E-08	5.4E-08
Sm-147	1.06E+11 y	0.005	1.4E-06	5.0E-04	1.4E-07	9.2E-08	6.4E-08	5.2E-08	4.9E-08
Sm-151	90.0 y	0.005	1.5E-09	5.0E-04	6.4E-10	3.3E-10	2.0E-10	1.2E-10	9.8E-11
Sm-153	1.95 d	0.005	8.4E-09	5.0E-04	5.4E-09	2.7E-09	1.6E-09	9.2E-10	7.4E-10

Table A.1.—*(continued)*

Nuclide	Physical half-life	f_1 <1y	$e(\tau)$ 3 months	f_1 ≥1y	$e(\tau)$ 1 Year	5 Years	10 Years	15 Years	Adult
Sm-155	0.368 h	0.005	3.6E-10	5.0E-04	2.0E-10	9.7E-11	5.5E-11	3.7E-11	2.9E-11
Sm-156	9.40 h	0.005	2.8E-09	5.0E-04	1.8E-09	9.0E-10	5.4E-10	3.1E-10	2.5E-10
Europium									
Eu-145	5.94 d	0.005	5.1E-09	5.0E-04	3.7E-09	2.1E-09	1.4E-09	9.4E-10	7.5E-10
Eu-146	4.61 d	0.005	8.5E-09	5.0E-04	6.2E-09	3.6E-09	2.4E-09	1.6E-09	1.3E-09
Eu-147	24.0 d	0.005	3.7E-09	5.0E-04	2.5E-09	1.4E-09	8.9E-10	5.6E-10	4.4E-10
Eu-148	54.5 d	0.005	8.5E-09	5.0E-04	6.0E-09	3.5E-09	2.4E-09	1.6E-09	1.3E-09
Eu-149	93.1 d	0.005	9.7E-10	5.0E-04	6.3E-10	3.4E-10	2.1E-10	1.3E-10	1.0E-10
Eu-150	34.2 y	0.005	1.3E-08	5.0E-04	5.7E-09	3.4E-09	2.3E-09	1.5E-09	1.3E-09
Eu-150	12.6 h	0.005	4.4E-09	5.0E-04	2.8E-09	1.4E-09	8.2E-10	4.7E-10	3.8E-10
Eu-152	13.3 y	0.005	1.6E-08	5.0E-04	7.4E-09	4.1E-09	2.6E-09	1.7E-09	1.4E-09
Eu-152m	9.32 h	0.005	5.7E-09	5.0E-04	3.6E-09	1.8E-09	1.1E-09	6.2E-10	5.0E-10
Eu-154	8.80 y	0.005	2.5E-08	5.0E-04	1.2E-08	6.5E-09	4.1E-09	2.5E-09	2.0E-09
Eu-155	4.96 y	0.005	4.3E-09	5.0E-04	2.2E-09	1.1E-09	6.8E-10	4.0E-10	3.2E-10
Eu-156	15.2 d	0.005	2.2E-08	5.0E-04	1.5E-08	7.5E-09	4.6E-09	2.7E-09	2.2E-09
Eu-157	15.1 h	0.005	6.7E-09	5.0E-04	4.3E-09	2.2E-09	1.3E-09	7.5E-10	6.0E-10
Eu-158	0.765 h	0.005	1.1E-09	5.0E-04	6.2E-10	3.1E-10	1.8E-10	1.2E-10	9.4E-11
Gadolinium									
Gd-145	0.382 h	0.005	4.5E-10	5.0E-04	2.6E-10	1.3E-10	8.1E-11	5.6E-11	4.4E-11
Gd-146	48.3 d	0.005	9.4E-09	5.0E-04	6.0E-09	3.2E-09	2.0E-09	1.2E-09	9.6E-10
Gd-147	1.59 d	0.005	4.5E-09	5.0E-04	3.2E-09	1.8E-09	1.2E-09	7.7E-10	6.1E-10
Gd-148	93.0 y	0.005	1.7E-06	5.0E-04	1.6E-07	1.1E-07	7.3E-08	5.9E-08	5.6E-08
Gd-149	9.40 d	0.005	4.0E-09	5.0E-04	2.7E-09	1.5E-09	9.3E-10	5.7E-10	4.5E-10
Gd-151	120 d	0.005	2.1E-09	5.0E-04	1.3E-09	6.8E-10	4.2E-10	2.4E-10	2.0E-10
Gd-152	1.08E+14 y	0.005	1.2E-06	5.0E-04	1.2E-07	7.7E-08	5.3E-08	4.3E-08	4.1E-08
Gd-153	242 d	0.005	2.9E-09	5.0E-04	1.8E-09	9.4E-10	5.8E-10	3.4E-10	2.7E-10
Gd-159	18.6 h	0.005	5.7E-09	5.0E-04	3.6E-09	1.8E-09	1.1E-09	6.2E-10	4.9E-10
Terbium									
Tb-147	1.65 h	0.005	1.5E-09	5.0E-04	1.0E-09	5.4E-10	3.3E-10	2.0E-10	1.6E-10
Tb-149	4.15 h	0.005	2.4E-09	5.0E-04	1.5E-09	8.0E-10	5.0E-10	3.1E-10	2.5E-10

Table A.1.—(continued)

Nuclide	Physical half-life	f_1 <1y	e(τ) 3 months	f_1 ≥1y	e(τ) 1 Year	5 Years	10 Years	15 Years	Adult
Tb-150	3.27 h	0.005	2.5E-09	5.0E-04	1.6E-09	8.3E-10	5.1E-10	3.2E-10	2.5E-10
Tb-151	17.6 h	0.005	2.7E-09	5.0E-04	1.9E-09	1.0E-09	6.7E-10	4.2E-10	3.4E-10
Tb-153	2.34 d	0.005	2.3E-09	5.0E-04	1.5E-09	8.2E-10	5.1E-10	3.1E-10	2.5E-10
Tb-154	21.4 h	0.005	4.7E-09	5.0E-04	3.4E-09	1.9E-09	1.3E-09	8.1E-10	6.5E-10
Tb-155	5.32 d	0.005	1.9E-09	5.0E-04	1.3E-09	6.8E-10	4.3E-10	2.6E-10	2.1E-10
Tb-156	5.34 d	0.005	9.0E-09	5.0E-04	6.3E-09	3.5E-09	2.3E-09	1.5E-09	1.2E-09
Tb-156m	1.02 d	0.005	1.5E-09	5.0E-04	1.0E-09	5.6E-10	3.5E-10	2.2E-10	1.7E-10
Tb-156m	5.00 h	0.005	8.0E-10	5.0E-04	5.2E-10	2.7E-10	1.7E-10	1.0E-10	8.1E-11
Tb-157	1.50E+02 y	0.005	4.9E-10	5.0E-04	2.2E-10	1.1E-10	6.8E-11	4.1E-11	3.4E-11
Tb-158	1.50E+02 y	0.005	1.3E-08	5.0E-04	5.9E-09	3.3E-09	2.1E-09	1.4E-09	1.1E-09
Tb-160	72.3 d	0.005	1.6E-08	5.0E-04	1.0E-08	5.4E-09	3.3E-09	2.0E-09	1.6E-09
Tb-161	6.91 d	0.005	8.3E-09	5.0E-04	5.3E-09	2.7E-09	1.6E-09	9.0E-10	7.2E-10
Dysprosium									
Dy-155	10.0 h	0.005	9.7E-10	5.0E-04	6.8E-10	3.8E-10	2.5E-10	1.6E-10	1.3E-10
Dy-157	8.10 h	0.005	4.4E-10	5.0E-04	3.1E-10	1.8E-10	1.2E-10	7.7E-11	6.1E-11
Dy-159	144 d	0.005	1.0E-09	5.0E-04	6.4E-10	3.4E-10	2.1E-10	1.3E-10	1.0E-10
Dy-165	2.33 h	0.005	1.3E-09	5.0E-04	7.9E-10	3.9E-10	2.3E-10	1.4E-10	1.1E-10
Dy-166	3.40 d	0.005	1.9E-08	5.0E-04	1.2E-08	6.0E-09	3.6E-09	2.0E-09	1.6E-09
Holmium									
Ho-155	0.800 h	0.005	3.8E-10	5.0E-04	2.3E-10	1.2E-10	7.1E-11	4.7E-11	3.7E-11
Ho-157	0.210 h	0.005	5.8E-11	5.0E-04	3.6E-11	1.9E-11	1.2E-11	8.1E-12	6.5E-12
Ho-159	0.550 h	0.005	7.1E-11	5.0E-04	4.3E-11	2.3E-11	1.4E-11	9.9E-12	7.9E-12
Ho-161	2.50 h	0.005	1.4E-10	5.0E-04	8.1E-11	4.2E-11	2.5E-11	1.6E-11	1.3E-11
Ho-162	0.250 h	0.005	3.5E-11	5.0E-04	2.0E-11	1.0E-11	6.0E-12	4.2E-12	3.3E-12
Ho-162m	1.13 h	0.005	2.4E-10	5.0E-04	1.5E-10	7.9E-11	4.9E-11	3.3E-11	2.6E-11
Ho-164	0.483 h	0.005	1.2E-10	5.0E-04	6.5E-11	3.2E-11	1.8E-11	1.2E-11	9.5E-12
Ho-164m	0.625 h	0.005	2.0E-10	5.0E-04	1.1E-10	5.5E-11	3.2E-11	2.1E-11	1.6E-11
Ho-166	1.12 d	0.005	1.6E-08	5.0E-04	1.0E-08	5.2E-09	3.1E-09	1.7E-09	1.4E-09
Ho-166m	1.20E+03 y	0.005	2.6E-08	5.0E-04	9.3E-09	5.3E-09	3.5E-09	2.4E-09	2.0E-09
Ho-167	3.10 h	0.005	8.8E-10	5.0E-04	5.5E-10	2.8E-10	1.7E-10	1.0E-10	8.3E-11

Table A.1.—(continued)

Nuclide	Physical half-life	f_1 <1y	$e(\tau)$ 3 months	f_1 ≥1y	$e(\tau)$ 1 Year	5 Years	10 Years	15 Years	Adult
Erbium									
Er-161	3.24 h	0.005	6.5E-10	5.0E-04	4.4E-10	2.4E-10	1.6E-10	1.0E-10	8.0E-11
Er-165	10.4 h	0.005	1.7E-10	5.0E-04	1.1E-10	6.2E-11	3.9E-11	2.4E-11	1.9E-11
Er-169	9.30 d	0.005	4.4E-09	5.0E-04	2.8E-09	1.4E-09	8.2E-10	4.7E-10	3.7E-10
Er-171	7.52 h	0.005	4.0E-09	5.0E-04	2.5E-09	1.3E-09	7.6E-10	4.5E-10	3.6E-10
Er-172	2.05 d	0.005	1.0E-08	5.0E-04	6.8E-09	3.5E-09	2.1E-09	1.3E-09	1.0E-09
Thulium									
Tm-162	0.362 h	0.005	2.9E-10	5.0E-04	1.7E-10	8.7E-11	5.2E-11	3.6E-11	2.9E-11
Tm-166	7.70 h	0.005	2.1E-09	5.0E-04	1.5E-09	8.3E-10	5.5E-10	3.5E-10	2.8E-10
Tm-167	9.24 d	0.005	6.0E-09	5.0E-04	3.9E-09	2.0E-09	1.2E-09	7.0E-10	5.6E-10
Tm-170	129 d	0.005	1.6E-08	5.0E-04	9.8E-09	4.9E-09	2.9E-09	1.6E-09	1.3E-09
Tm-171	1.92 y	0.005	1.5E-09	5.0E-04	7.8E-10	3.9E-10	2.3E-10	1.3E-10	1.1E-10
Tm-172	2.65 d	0.005	1.9E-08	5.0E-04	1.2E-08	6.1E-09	3.7E-09	2.1E-09	1.7E-09
Tm-173	8.24 h	0.005	3.3E-09	5.0E-04	2.1E-09	1.1E-09	6.5E-10	3.8E-10	3.1E-10
Tm-175	0.253 h	0.005	3.1E-10	5.0E-04	1.7E-10	8.6E-11	5.0E-11	3.4E-11	2.7E-11
Ytterbium									
Yb-162	0.315 h	0.005	2.2E-10	5.0E-04	1.3E-10	6.9E-11	4.2E-11	2.9E-11	2.3E-11
Yb-166	2.36 d	0.005	7.7E-09	5.0E-04	5.4E-09	2.9E-09	1.9E-09	1.2E-09	9.5E-10
Yb-167	0.292 h	0.005	7.0E-11	5.0E-04	4.1E-11	2.1E-11	1.2E-11	8.4E-12	6.7E-12
Yb-169	32.0 d	0.005	7.1E-09	5.0E-04	4.6E-09	2.4E-09	1.5E-09	8.8E-10	7.1E-10
Yb-175	4.19 d	0.005	5.0E-09	5.0E-04	3.2E-09	1.6E-09	9.5E-10	5.4E-10	4.4E-10
Yb-177	1.90 h	0.005	1.0E-09	5.0E-04	6.8E-10	3.4E-10	2.0E-10	1.1E-10	8.8E-11
Yb-178	1.23 h	0.005	1.4E-09	5.0E-04	8.4E-10	4.2E-10	2.4E-10	1.5E-10	1.2E-10
Lutetium									
Lu-169	1.42 d	0.005	3.5E-09	5.0E-04	2.4E-09	1.4E-09	8.9E-10	5.7E-10	4.6E-10
Lu-170	2.00 d	0.005	7.4E-09	5.0E-04	5.2E-09	2.9E-09	1.9E-09	1.2E-09	9.9E-10
Lu-171	8.22 d	0.005	5.9E-09	5.0E-04	4.0E-09	2.2E-09	1.4E-09	8.5E-10	6.7E-10
Lu-172	6.70 d	0.005	1.0E-08	5.0E-04	7.0E-09	3.9E-09	2.5E-09	1.6E-09	1.3E-09
Lu-173	1.37 y	0.005	2.7E-09	5.0E-04	1.6E-09	8.6E-10	5.3E-10	3.2E-10	2.6E-10
Lu-174	3.31 y	0.005	3.2E-09	5.0E-04	1.7E-09	9.1E-10	5.6E-10	3.3E-10	2.7E-10

Table A.1.—(continued)

Nuclide	Physical half-life	f_1 <1y	$e(\tau)$ 3 months	f_1 ≥1y	$e(\tau)$ 1 Year	5 Years	10 Years	15 Years	Adult
Lu-174m	142 d	0.005	6.2E-09	5.0E-04	3.8E-09	1.9E-09	1.1E-09	6.6E-10	5.3E-10
Lu-176	3.60E+10 y	0.005	2.4E-08	5.0E-04	1.1E-08	5.7E-09	3.5E-09	2.2E-09	1.8E-09
Lu-176m	3.68 h	0.005	2.0E-09	5.0E-04	1.2E-09	6.0E-10	3.5E-10	2.1E-10	1.7E-10
Lu-177	6.71 d	0.005	6.1E-09	5.0E-04	3.9E-09	2.0E-09	1.2E-09	6.6E-10	5.3E-10
Lu-177m	161 d	0.005	1.7E-08	5.0E-04	1.1E-08	5.8E-09	3.6E-09	2.1E-09	1.7E-09
Lu-178	0.473 h	0.005	5.9E-10	5.0E-04	3.3E-10	1.6E-10	9.0E-11	6.1E-11	4.7E-11
Lu-178m	0.378 h	0.005	4.3E-10	5.0E-04	2.4E-10	1.2E-10	7.1E-11	4.9E-11	3.8E-11
Lu-179	4.59 h	0.005	2.4E-09	5.0E-04	1.5E-09	7.5E-10	4.4E-10	2.6E-10	2.1E-10
Hafnium									
Hf-170	16.0 h	0.020	3.9E-09	0.002	2.7E-09	1.5E-09	9.5E-10	6.0E-10	4.8E-10
Hf-172	1.87 y	0.020	1.9E-08	0.002	6.1E-09	3.3E-09	2.0E-09	1.3E-09	1.0E-09
Hf-173	24.0 h	0.020	1.9E-09	0.002	1.3E-09	7.2E-10	4.6E-10	2.8E-10	2.3E-10
Hf-175	70.0 d	0.020	3.8E-09	0.002	2.4E-09	1.3E-09	8.4E-10	5.2E-10	4.1E-10
Hf-177m	0.856 h	0.020	7.8E-10	0.002	4.7E-10	2.5E-10	1.5E-10	1.0E-10	9.1E-11
Hf-178m	31.0 y	0.020	7.0E-08	0.002	1.9E-08	1.1E-08	7.8E-09	5.5E-09	4.7E-09
Hf-179m	25.1 d	0.020	1.2E-08	0.002	7.8E-09	4.1E-09	2.6E-09	1.6E-09	1.2E-09
Hf-180m	5.50 h	0.020	1.4E-09	0.002	9.7E-10	5.3E-10	3.3E-10	2.1E-10	1.7E-10
Hf-181	42.4 d	0.020	1.2E-08	0.002	7.4E-09	3.8E-09	2.3E-09	1.4E-09	1.1E-09
Hf-182	9.00E+06 y	0.020	5.6E-08	0.002	7.9E-09	5.4E-09	4.0E-09	3.3E-09	3.0E-09
Hf-182m	1.02 h	0.020	4.1E-10	0.002	2.5E-10	1.3E-10	7.8E-11	5.2E-11	4.2E-11
Hf-183	1.07 h	0.020	8.1E-10	0.002	4.8E-10	2.4E-10	1.4E-10	9.3E-11	7.3E-11
Hf-184	4.12 h	0.020	5.5E-09	0.002	3.6E-09	1.8E-09	1.1E-09	6.6E-10	5.2E-10
Tantalum									
Ta-172	0.613 h	0.010	5.5E-10	0.001	3.2E-10	1.6E-10	9.8E-11	6.6E-11	5.3E-11
Ta-173	3.65 h	0.010	2.0E-09	0.001	1.3E-09	6.5E-10	3.9E-10	2.4E-10	1.9E-10
Ta-174	1.20 h	0.010	6.2E-10	0.001	3.7E-10	1.9E-10	1.1E-10	7.2E-11	5.7E-11
Ta-175	10.5 h	0.010	1.6E-09	0.001	1.1E-09	6.2E-10	4.0E-10	2.6E-10	2.1E-10
Ta-176	8.08 h	0.010	2.4E-09	0.001	1.7E-09	9.2E-10	6.1E-10	3.9E-10	3.1E-10
Ta-177	2.36 d	0.010	1.0E-09	0.001	6.9E-10	3.6E-10	2.2E-10	1.3E-10	1.1E-10
Ta-178	2.20 h	0.010	6.3E-10	0.001	4.5E-10	2.4E-10	1.5E-10	9.1E-11	7.2E-11

Table A.1.—(continued)

Nuclide	Physical half-life	f_1 <1y	$e(\tau)$ 3 months	f_1 ≥1y	$e(\tau)$ 1 Year	5 Years	10 Years	15 Years	Adult
Ta-179	1.82 y	0.010	6.2E-10	0.001	4.1E-10	2.2E-10	1.3E-10	8.1E-11	6.5E-11
Ta-180	1.00E+13 y	0.010	8.1E-09	0.001	5.3E-09	2.8E-09	1.7E-09	1.1E-09	8.4E-10
Ta-180m	8.10 h	0.010	5.8E-10	0.001	3.7E-10	1.9E-10	1.1E-10	6.7E-11	5.4E-11
Ta-182	115 d	0.010	1.4E-08	0.001	9.4E-09	5.0E-09	3.1E-09	1.9E-09	1.5E-09
Ta-182m	0.264 h	0.010	1.4E-10	0.001	7.5E-11	3.7E-11	2.1E-11	1.5E-11	1.2E-11
Ta-183	5.10 d	0.010	1.4E-08	0.001	9.3E-09	4.7E-09	2.8E-09	1.6E-09	1.3E-09
Ta-184	8.70 h	0.010	6.7E-09	0.001	4.4E-09	2.3E-09	1.4E-09	8.5E-10	6.8E-10
Ta-185	0.816 h	0.010	8.3E-10	0.001	4.6E-10	2.3E-10	1.3E-10	8.6E-11	6.8E-11
Ta-186	0.175 h	0.010	3.8E-10	0.001	2.1E-10	1.1E-10	6.1E-11	4.2E-11	3.3E-11
Tungsten									
W-176	2.30 h	0.600	6.8E-10	0.300	5.5E-10	3.0E-10	2.0E-10	1.3E-10	1.0E-10
W-177	2.25 h	0.600	4.4E-10	0.300	3.2E-10	1.7E-10	1.1E-10	7.2E-11	5.8E-11
W-178	21.7 d	0.600	1.8E-09	0.300	1.4E-09	7.3E-10	4.5E-10	2.7E-10	2.2E-10
W-179	0.625 h	0.600	3.4E-11	0.300	2.0E-11	1.0E-11	6.2E-12	4.2E-12	3.3E-12
W-181	121 d	0.600	6.3E-10	0.300	4.7E-10	2.5E-10	1.6E-10	9.5E-11	7.6E-11
W-185	75.1 d	0.600	4.4E-09	0.300	3.3E-09	1.6E-09	9.7E-10	5.5E-10	4.4E-10
W-187	23.9 h	0.600	5.5E-09	0.300	4.3E-09	2.2E-09	1.3E-09	7.8E-10	6.3E-10
W-188	69.4 d	0.600	2.1E-08	0.300	1.5E-08	7.7E-09	4.6E-09	2.6E-09	2.1E-09
Rhenium									
Re-177	0.233 h	1.000	2.5E-10	0.800	1.4E-10	7.2E-11	4.1E-11	2.8E-11	2.2E-11
Re-178	0.220 h	1.000	2.9E-10	0.800	1.6E-10	7.9E-11	4.6E-11	3.1E-11	2.5E-11
Re-181	20.0 h	1.000	4.2E-09	0.800	2.8E-09	1.4E-09	8.2E-10	5.4E-10	4.2E-10
Re-182	2.67 d	1.000	1.4E-08	0.800	8.9E-09	4.7E-09	2.8E-09	1.8E-09	1.4E-09
Re-182	12.7 h	1.000	2.4E-09	0.800	1.7E-09	8.9E-10	5.2E-10	3.5E-10	2.7E-10
Re-184	38.0 d	1.000	8.9E-09	0.800	5.6E-09	3.0E-09	1.8E-09	1.3E-09	1.0E-09
Re-184m	165 d	1.000	1.7E-08	0.800	9.8E-09	4.9E-09	2.8E-09	1.9E-09	1.5E-09
Re-186	3.78 d	1.000	1.9E-08	0.800	1.1E-08	5.5E-09	3.0E-09	1.9E-09	1.5E-09
Re-186m	2.00E+05 y	1.000	3.0E-08	0.800	1.6E-08	7.6E-09	4.4E-09	2.8E-09	2.2E-09
Re-187	5.00E+10 y	1.000	6.8E-11	0.800	3.8E-11	1.8E-11	1.0E-11	6.6E-12	5.1E-12
Re-188	17.0 h	1.000	1.7E-08	0.800	1.1E-08	5.4E-09	2.9E-09	1.8E-09	1.4E-09

Table A.1.—(continued)

Nuclide	Physical half-life	f_1 <1y	e(τ) 3 months	f_1 ≥1y	e(τ) 1 Year	5 Years	10 Years	15 Years	Adult
Fe-188m	0.310 h	1.000	3.8E-10	0.800	2.3E-10	1.1E-10	6.1E-11	4.0E-11	3.0E-11
Fe-189	1.01 d	1.000	9.8E-09	0.800	6.2E-09	3.0E-09	1.6E-09	1.0E-09	7.3E-10
Osmium									
Os-180	0.366 h	0.020	1.6E-10	0.010	9.8E-11	5.1E-11	3.2E-11	2.2E-11	1.7E-11
Os-181	1.75 h	0.020	7.6E-10	0.010	5.0E-10	2.7E-10	1.7E-10	1.1E-10	8.9E-11
Os-182	22.0 h	0.020	4.6E-09	0.010	3.2E-09	1.7E-09	1.1E-09	7.0E-10	5.6E-10
Os-185	94.0 d	0.020	3.8E-09	0.010	2.6E-09	1.5E-09	9.8E-10	6.5E-10	5.1E-10
Os-189m	6.00 h	0.020	2.1E-10	0.010	1.3E-10	6.5E-11	3.8E-11	2.2E-11	1.8E-11
Os-191	15.4 d	0.020	6.3E-09	0.010	4.1E-09	2.1E-09	1.2E-09	7.0E-10	5.7E-10
Os-191m	13.0 h	0.020	1.1E-09	0.010	7.1E-10	3.5E-10	2.1E-10	1.2E-10	9.6E-11
Os-193	1.25 d	0.020	9.3E-09	0.010	6.0E-09	3.0E-09	1.8E-09	1.0E-09	8.1E-10
Os-194	6.00 y	0.020	2.9E-08	0.010	1.7E-08	8.8E-09	5.2E-09	3.0E-09	2.4E-09
Iridium									
Ir-182	0.250 h	0.020	5.3E-10	0.010	3.0E-10	1.5E-10	8.9E-11	6.0E-11	4.8E-11
Ir-184	3.02 h	0.020	1.5E-09	0.010	9.7E-10	5.2E-10	3.3E-10	2.1E-10	1.7E-10
Ir-185	14.0 h	0.020	2.4E-09	0.010	1.6E-09	8.6E-10	5.3E-10	3.3E-10	2.6E-10
Ir-186	15.8 h	0.020	3.8E-09	0.010	2.7E-09	1.5E-09	9.6E-10	6.1E-10	4.9E-10
Ir-186	1.75 h	0.020	5.8E-10	0.010	3.6E-10	2.1E-10	1.3E-10	7.7E-11	6.1E-11
Ir-187	10.5 h	0.020	1.1E-09	0.010	7.3E-10	3.9E-10	2.5E-10	1.5E-10	1.2E-10
Ir-188	1.73 d	0.020	4.6E-09	0.010	3.3E-09	1.8E-09	1.2E-09	7.9E-10	6.3E-10
Ir-189	13.3 d	0.020	2.5E-09	0.010	1.7E-09	8.6E-10	5.2E-10	3.0E-10	2.4E-10
Ir-190	12.1 d	0.020	1.0E-08	0.010	7.1E-09	3.9E-09	2.5E-09	1.6E-09	1.2E-09
Ir-190m	3.10 h	0.020	9.4E-10	0.010	6.4E-10	3.5E-10	2.3E-10	1.5E-10	1.2E-10
Ir-190m	1.20 h	0.020	7.9E-11	0.010	5.0E-11	2.6E-11	1.6E-11	1.0E-11	8.0E-12
Ir-192	74.0 d	0.020	1.3E-08	0.010	8.7E-09	4.6E-09	2.8E-09	1.7E-09	1.4E-09
Ir-192m	2.41E+02 y	0.020	2.8E-09	0.010	1.4E-09	8.3E-10	5.5E-10	3.7E-10	3.1E-10
Ir-193m	11.9 d	0.020	3.2E-09	0.010	2.0E-09	1.0E-09	6.0E-10	3.4E-10	2.7E-10
Ir-194	19.1 h	0.020	1.5E-08	0.010	9.8E-09	4.9E-09	2.9E-09	1.7E-09	1.3E-09
Ir-194m	171 d	0.020	1.7E-08	0.010	1.1E-08	6.4E-09	4.1E-09	2.6E-09	2.1E-09
Ir-195	2.50 h	0.020	1.2E-09	0.010	7.3E-10	3.6E-10	2.1E-10	1.3E-10	1.0E-10

Table A.1.—(*continued*)

Nuclide	Physical half-life	f_1 <1y	e(τ) 3 months	f_1 ≥1y	e(τ) 1 Year	5 Years	10 Years	15 Years	Adult
Ir-195m	3.80 h	0.020	2.3E-09	0.010	1.5E-09	7.3E-10	4.3E-10	2.6E-10	2.1E-10
Platinum									
Pt-186	2.00 h	0.020	7.8E-10	0.010	5.3E-10	2.9E-10	1.8E-10	1.2E-10	9.3E-11
Pt-188	10.2 d	0.020	6.7E-09	0.010	4.5E-09	2.4E-09	1.5E-09	9.5E-10	7.6E-10
Pt-189	10.9 h	0.020	1.1E-09	0.010	7.4E-10	3.9E-10	2.5E-10	1.5E-10	1.2E-10
Pt-191	2.80 d	0.020	3.1E-09	0.010	2.1E-09	1.1E-09	6.9E-10	4.2E-10	3.4E-10
Pt-193	50.0 y	0.020	3.7E-10	0.010	2.4E-10	1.2E-10	6.9E-11	3.9E-11	3.1E-11
Pt-193m	4.33 d	0.020	5.2E-09	0.010	3.4E-09	1.7E-09	9.9E-10	5.6E-10	4.5E-10
Pt-195m	4.02 d	0.020	7.1E-09	0.010	4.6E-09	2.3E-09	1.4E-09	7.9E-10	6.3E-10
Pt-197	18.3 h	0.020	4.7E-09	0.010	3.0E-09	1.5E-09	8.8E-10	5.1E-10	4.0E-10
Pt-197m	1.57 h	0.020	1.0E-09	0.010	6.1E-10	3.0E-10	1.8E-10	1.1E-10	8.4E-11
Pt-199	0.513 h	0.020	4.7E-10	0.010	2.7E-10	1.3E-10	7.5E-11	5.0E-11	3.9E-11
Pt-200	12.5 h	0.020	1.4E-08	0.010	8.8E-09	4.4E-09	2.6E-09	1.5E-09	1.2E-09
Gold									
Au-193	17.6 h	0.200	1.2E-09	0.100	8.8E-10	4.6E-10	2.8E-10	1.7E-10	1.3E-10
Au-194	1.65 d	0.200	2.9E-09	0.100	2.2E-09	1.2E-09	8.1E-10	5.3E-10	4.2E-10
Au-195	183 d	0.200	2.4E-09	0.100	1.7E-09	8.9E-10	5.4E-10	3.2E-10	2.5E-10
Au-198	2.69 d	0.200	1.0E-08	0.100	7.2E-09	3.7E-09	2.2E-09	1.3E-09	1.0E-09
Au-198m	2.30 d	0.200	1.2E-08	0.100	8.5E-09	4.4E-09	2.7E-09	1.6E-09	1.3E-09
Au-199	3.14 d	0.200	4.5E-09	0.100	3.1E-09	1.6E-09	9.5E-10	5.5E-10	4.4E-10
Au-200	0.807 h	0.200	8.3E-10	0.100	4.7E-10	2.3E-10	1.3E-10	8.7E-11	6.8E-11
Au-200m	18.7 h	0.200	9.2E-09	0.100	6.6E-09	3.5E-09	2.2E-09	1.3E-09	1.1E-09
Au-201	0.440 h	0.200	3.1E-10	0.100	1.7E-10	8.2E-11	4.6E-11	3.1E-11	2.4E-11
Mercury									
Hg-193 (organic)	3.50 h	1.000 0.800	3.3E-10 4.7E-10	1.000 0.400	1.9E-10 4.4E-10	9.8E-11 2.2E-10	5.8E-11 1.4E-10	3.9E-11 8.3E-11	3.1E-11 6.6E-11
Hg-193 (inorganic)	3.50 h	0.040	8.5E-10	0.020	5.5E-10	2.8E-10	1.7E-10	1.0E-10	8.2E-11
Hg-193m (organic)	11.1 h	1.000 0.800	1.1E-09 1.6E-09	1.000 0.400	6.8E-10 1.8E-09	3.7E-10 9.5E-10	2.3E-10 6.0E-10	1.5E-10 3.7E-10	1.3E-10 3.0E-10
Hg-193m (inorganic)	11.1 h	0.040	3.6E-09	0.020	2.4E-09	1.3E-09	8.1E-10	5.0E-10	4.0E-10

Table A.1.—(continued)

Nuclide	Physical half-life	f_1 <1y	e(τ) 3 months	f_1 ≥1y	e(τ) 1 Year	5 Years	10 Years	15 Years	Adult
Hg-194 (organic)	2.60E+02 y	1.000 0.800	1.3E-07 1.1E-07	1.000 0.400	1.2E-07 4.8E-08	8.4E-08 3.5E-08	6.6E-08 2.7E-08	5.5E-08 2.3E-08	5.1E-08 2.1E-08
Hg-194 (inorganic)	2.60E+02 y	0.040	7.2E-09	0.020	3.6E-09	2.6E-09	1.9E-09	1.5E-09	1.4E-09
Hg-195 (organic)	9.90 h	1.000 0.800	3.0E-10 4.6E-10	1.000 0.400	2.0E-10 4.8E-10	1.0E-10 2.5E-10	6.4E-11 1.5E-10	4.2E-11 9.3E-11	3.4E-11 7.5E-11
Hg-195 (inorganic)	9.90 h	0.040	9.5E-10	0.020	6.3E-10	3.3E-10	2.0E-10	1.2E-10	9.7E-11
Hg-195m (organic)	1.73 d	1.000 0.800	2.1E-09 2.6E-09	1.000 0.400	1.3E-09 2.8E-09	6.8E-10 1.4E-09	4.2E-10 8.7E-10	2.7E-10 5.1E-10	2.2E-10 4.1E-10
Hg-195m (inorganic)	1.73 d	0.040	5.8E-09	0.020	3.8E-09	2.0E-09	1.2E-09	7.0E-10	5.6E-10
Hg-197 (organic)	2.67 d	1.000 0.800	9.7E-10 1.3E-09	1.000 0.400	6.2E-10 1.2E-09	3.1E-10 6.1E-10	1.9E-10 3.7E-10	1.2E-10 2.2E-10	9.9E-11 1.7E-10
Hg-197 (inorganic)	2.67 d	0.040	2.5E-09	0.020	1.6E-09	8.3E-10	5.0E-10	2.9E-10	2.3E-10
Hg-197m (organic)	23.8 h	1.000 0.800	1.5E-09 2.2E-09	1.000 0.400	9.5E-10 2.5E-09	4.8E-10 1.2E-09	2.9E-10 7.3E-10	1.8E-10 4.2E-10	1.5E-10 3.4E-10
Hg-197m (inorganic)	23.8 h	0.040	5.2E-09	0.020	3.4E-09	1.7E-09	1.0E-09	5.9E-10	4.7E-10
Hg-199m (organic)	0.710 h	1.000 0.800	3.4E-10 3.6E-10	1.000 0.400	1.9E-10 2.1E-10	9.3E-11 1.0E-10	5.3E-11 5.8E-11	3.6E-11 3.9E-11	2.8E-11 3.1E-11
Hg-199m (inorganic)	0.710 h	0.040	3.7E-10	0.020	2.1E-10	1.0E-10	5.9E-11	3.9E-11	3.1E-11
Hg-203 (organic)	46.6 d	1.000 0.800	1.5E-08 1.3E-08	1.000 0.400	1.1E-08 6.4E-09	5.7E-09 3.4E-09	3.6E-09 2.1E-09	2.3E-09 1.3E-09	1.9E-09 1.1E-09
Hg-203 (inorganic)	46.6 d	0.040	5.5E-09	0.020	3.6E-09	1.8E-09	1.1E-09	6.7E-10	5.4E-10
Thallium									
Tl-194	0.550 h	1.000	6.1E-11	1.000	3.9E-11	2.2E-11	1.4E-11	1.0E-11	8.1E-12
Tl-194m	0.546 h	1.000	3.8E-10	1.000	2.2E-10	1.2E-10	7.0E-11	4.9E-11	4.0E-11
Tl-195	1.16 h	1.000	2.3E-10	1.000	1.4E-10	7.5E-11	4.7E-11	3.3E-11	2.7E-11
Tl-197	2.84 h	1.000	2.1E-10	1.000	1.3E-10	6.7E-11	4.2E-11	2.8E-11	2.3E-11
Tl-198	5.30 h	1.000	4.7E-10	1.000	3.3E-10	1.9E-10	1.2E-10	8.7E-11	7.3E-11
Tl-198m	1.87 h	1.000	4.8E-10	1.000	3.0E-10	1.6E-10	9.7E-11	6.7E-11	5.4E-11
Tl-199	7.42 h	1.000	2.3E-10	1.000	1.5E-10	7.7E-11	4.8E-11	3.2E-11	2.6E-11
Tl-200	1.09 d	1.000	1.3E-09	1.000	9.1E-10	5.3E-10	3.5E-10	2.4E-10	2.0E-10

Table A.1.—(continued)

Nuclide	Physical half-life	f_1 <1y	e(τ) 3 months	f_1 ≥1y	e(τ) 1 Year	5 Years	10 Years	15 Years	Adult
Tl-201	3.04 d	1.000	8.4E-10	1.000	5.5E-10	2.9E-10	1.8E-10	1.2E-10	9.5E-11
Tl-202	12.2 d	1.000	2.9E-09	1.000	2.1E-09	1.2E-09	7.9E-10	5.4E-10	4.5E-10
Tl-204	3.78 y	1.000	1.3E-08	1.000	8.5E-09	4.2E-09	2.5E-09	1.5E-09	1.2E-09
Lead[a,b]									
Pb-195m	0.263 h	0.600	2.6E-10	0.200	1.6E-10	8.4E-11	5.2E-11	3.5E-11	2.9E-11
Pb-198	2.40 h	0.600	5.9E-10	0.200	4.8E-10	2.7E-10	1.7E-10	1.1E-10	1.0E-10
Pb-199	1.50 h	0.600	3.5E-10	0.200	2.6E-10	1.5E-10	9.4E-11	6.3E-11	5.4E-11
Pb-200	21.5 h	0.600	2.5E-09	0.200	2.0E-09	1.1E-09	7.0E-10	4.4E-10	4.0E-10
Pb-201	9.40 h	0.600	9.4E-10	0.200	7.8E-10	4.3E-10	2.7E-10	1.8E-10	1.6E-10
Pb-202	3.00E+05 y	0.600	3.4E-08	0.200	1.6E-08	1.3E-08	1.9E-08	2.7E-08	8.8E-09
Pb-202m	3.62 h	0.600	7.6E-10	0.200	6.1E-10	3.5E-10	2.3E-10	1.5E-10	1.3E-10
Pb-203	2.17 d	0.600	1.6E-09	0.200	1.3E-09	6.8E-10	4.3E-10	2.7E-10	2.4E-10
Pb-205	1.43E+07 y	0.600	2.1E-09	0.200	9.9E-10	6.2E-10	6.1E-10	6.5E-10	2.8E-10
Pb-209	3.25 h	0.600	5.7E-10	0.200	3.8E-10	1.9E-10	1.1E-10	6.6E-11	5.7E-11
Pb-210	22.3 y	0.600	8.4E-06	0.200	3.6E-06	2.2E-06	1.9E-06	1.9E-06	6.9E-07
Pb-211	0.601 h	0.600	3.1E-09	0.200	1.4E-09	7.1E-10	4.1E-10	2.7E-10	1.8E-10
Pb-212	10.6 h	0.600	1.5E-07	0.200	6.3E-08	3.3E-08	2.0E-08	1.3E-08	6.0E-09
Pb-214	0.447 h	0.600	2.7E-09	0.200	1.0E-09	5.2E-10	3.1E-10	2.0E-10	1.4E-10
Bismuth									
Bi-200	0.606 h	0.100	4.2E-10	0.050	2.7E-10	1.5E-10	9.5E-11	6.4E-11	5.1E-11
Bi-201	1.80 h	0.100	1.0E-09	0.050	6.7E-10	3.6E-10	2.2E-10	1.4E-10	1.2E-10
Bi-202	1.67 h	0.100	6.4E-10	0.050	4.4E-10	2.5E-10	1.6E-10	1.1E-10	8.9E-11
Bi-203	11.8 h	0.100	3.5E-09	0.050	2.5E-09	1.4E-09	9.3E-10	6.0E-10	4.8E-10
Bi-205	15.3 d	0.100	6.1E-09	0.050	4.5E-09	2.6E-09	1.7E-09	1.1E-09	9.0E-10
Bi-206	6.24 d	0.100	1.4E-08	0.050	1.0E-08	5.7E-09	3.7E-09	2.4E-09	1.9E-09
Bi-207	38.0 y	0.100	1.0E-08	0.050	7.1E-09	3.9E-09	2.5E-09	1.6E-09	1.3E-09
Bi-210	5.01 d	0.100	1.5E-08	0.050	9.7E-09	4.8E-09	2.9E-09	1.6E-09	1.3E-09
Bi-210m	3.00E+06 y	0.100	2.1E-07	0.050	9.1E-08	4.7E-08	3.0E-08	1.9E-08	1.5E-08
Bi-212	1.01 h	0.100	3.2E-09	0.050	1.8E-09	8.7E-10	5.0E-10	3.3E-10	2.6E-10
Bi-213	0.761 h	0.100	2.5E-09	0.050	1.4E-09	6.7E-10	3.9E-10	2.5E-10	2.0E-10

a   Dose coefficients for this element are based on age-specific biokinetic data

b   The f_1 value for 1 to 15 year olds is 0.4

Table A.1.—(continued)

Nuclide	Physical half-life	f_1 <1y	$e(\tau)$ 3 months	f_1 ≥1y	$e(\tau)$ 1 Year	5 Years	10 Years	15 Years	Adult
Bi-214	0.332 h	0.100	1.4E-09	0.050	7.4E-10	3.6E-10	2.1E-10	1.4E-10	1.1E-10
Polonium[a]									
Po-203	0.612 h	1.000	2.9E-10	0.500	2.4E-10	1.3E-10	8.5E-11	5.8E-11	4.6E-11
Po-205	1.80 h	1.000	3.5E-10	0.500	2.8E-10	1.6E-10	1.1E-10	7.2E-11	5.8E-11
Po-207	5.83 h	1.000	4.4E-10	0.500	5.7E-10	3.2E-10	2.1E-10	1.4E-10	1.1E-10
Po-210	138 d	1.000	2.6E-05	0.500	8.8E-06	4.4E-06	2.6E-06	1.6E-06	1.2E-06
Astatine									
At-207	1.80 h	1.000	2.5E-09	1.000	1.6E-09	8.0E-10	4.8E-10	2.9E-10	2.4E-10
At-211	7.21 h	1.000	1.2E-07	1.000	7.8E-08	3.8E-08	2.3E-08	1.3E-08	1.1E-08
Francium									
Fr-222	0.240 h	1.000	6.2E-09	1.000	3.9E-09	2.0E-09	1.3E-09	8.5E-10	7.2E-10
Fr-223	0.363 h	1.000	2.6E-08	1.000	1.7E-08	8.3E-09	5.0E-09	2.9E-09	2.4E-09
Radium[a,b]									
Ra-223	11.4 d	0.600	5.3E-06	0.200	1.1E-06	5.7E-07	4.5E-07	3.7E-07	1.0E-07
Ra-224	3.66 d	0.600	2.7E-06	0.200	6.6E-07	3.5E-07	2.6E-07	2.0E-07	6.5E-08
Ra-225	14.8 d	0.600	7.1E-06	0.200	1.2E-06	6.1E-07	5.0E-07	4.4E-07	9.9E-08
Ra-226	1.60E+03 y	0.600	4.7E-06	0.200	9.6E-07	6.2E-07	8.0E-07	1.5E-06	2.8E-07
Ra-227	0.703 h	0.600	1.1E-09	0.200	4.3E-10	2.5E-10	1.7E-10	1.3E-10	8.1E-11
Ra-228	5.75 y	0.600	3.0E-05	0.200	5.7E-06	3.4E-06	3.9E-06	5.3E-06	6.9E-07
Actinium									
Ac-224	2.90 h	0.005	1.0E-08	5.0E-04	5.2E-09	2.6E-09	1.5E-09	8.8E-10	7.0E-10
Ac-225	10.0 d	0.005	4.6E-07	5.0E-04	1.8E-07	9.1E-08	5.4E-08	3.0E-08	2.4E-08
Ac-226	1.21 d	0.005	1.4E-07	5.0E-04	7.6E-08	3.8E-08	2.3E-08	1.3E-08	1.0E-08
Ac-227	21.8 y	0.005	3.3E-05	5.0E-04	3.1E-06	2.2E-06	1.5E-06	1.2E-06	1.1E-06
Ac-228	6.13 h	0.005	7.4E-09	5.0E-04	2.8E-09	1.4E-09	8.7E-10	5.3E-10	4.3E-10
Thorium[a]									
Th-226	0.515 h	0.005	4.4E-09	5.0E-04	2.4E-09	1.2E-09	6.7E-10	4.5E-10	3.5E-10
Th-227	18.7 d	0.005	3.0E-07	5.0E-04	7.0E-08	3.6E-08	2.3E-08	1.5E-08	8.8E-09
Th-228	1.91 y	0.005	3.7E-06	5.0E-04	3.7E-07	2.2E-07	1.5E-07	9.4E-08	7.2E-08
Th-229	7.34E+03 y	0.005	1.1E-05	5.0E-04	1.0E-06	7.8E-07	6.2E-07	5.3E-07	4.9E-07

a Dose coefficients for this element are based on age-specific biokinetic data

b The f_1 value for 1 to 15 year olds is 0.3

Table A.1.—(continued)

Nuclide	Physical half-life	f_1 <1y	$e(\tau)$ 3 months	f_1 ≥1y	$e(\tau)$ 1 Year	5 Years	10 Years	15 Years	Adult
Th-230	7.70E+04 y	0.005	4.1E-06	5.0E-04	4.1E-07	3.1E-07	2.4E-07	2.2E-07	2.1E-07
Th-231	1.06 d	0.005	3.9E-09	5.0E-04	2.5E-09	1.2E-09	7.4E-10	4.2E-10	3.4E-10
Th-232	1.40E+10 y	0.005	4.6E-06	5.0E-04	4.5E-07	3.5E-07	2.9E-07	2.5E-07	2.3E-07
Th-234	24.1 d	0.005	4.0E-08	5.0E-04	2.5E-08	1.3E-08	7.4E-09	4.2E-09	3.4E-09
Protactinium									
Pa-227	0.638 h	0.005	5.8E-09	5.0E-04	3.2E-09	1.5E-09	8.7E-10	5.8E-10	4.5E-10
Pa-228	22.0 h	0.005	1.2E-08	5.0E-04	4.8E-09	2.6E-09	1.6E-09	9.7E-10	7.8E-10
Pa-230	17.4 d	0.005	2.6E-08	5.0E-04	5.7E-09	3.1E-09	1.9E-09	1.1E-09	9.2E-10
Pa-231	3.27E+04 y	0.005	1.3E-05	5.0E-04	1.3E-06	1.1E-06	9.2E-07	8.0E-07	7.1E-07
Pa-232	1.31 d	0.005	6.3E-09	5.0E-04	4.2E-09	2.2E-09	1.4E-09	8.9E-10	7.2E-10
Pa-233	27.0 d	0.005	9.7E-09	5.0E-04	6.2E-09	3.2E-09	1.9E-09	1.1E-09	8.7E-10
Pa-234	6.70 h	0.005	5.0E-09	5.0E-04	3.2E-09	1.7E-09	1.0E-09	6.4E-10	5.1E-10
Uranium[a]									
U-230	20.8 d	0.040	7.9E-07	0.020	3.0E-07	1.5E-07	1.0E-07	6.6E-08	5.6E-08
U-231	4.20 d	0.040	3.1E-09	0.020	2.0E-09	1.0E-09	6.1E-10	3.5E-10	2.8E-10
U-232	72.0 y	0.040	2.5E-06	0.020	8.2E-07	5.8E-07	5.7E-07	6.4E-07	3.3E-07
U-233	1.58E+05 y	0.040	3.8E-07	0.020	1.4E-07	9.2E-08	7.8E-08	7.8E-08	5.1E-08
U-234	2.44E+05 y	0.040	3.7E-07	0.020	1.3E-07	8.8E-08	7.4E-08	7.4E-08	4.9E-08
U-235	7.04E+08 y	0.040	3.5E-07	0.020	1.3E-07	8.5E-08	7.1E-08	7.0E-08	4.7E-08
U-236	2.34E+07 y	0.040	3.5E-07	0.020	1.3E-07	8.4E-08	7.0E-08	7.0E-08	4.7E-08
U-237	6.75 d	0.040	8.3E-09	0.020	5.4E-09	2.8E-09	1.6E-09	9.5E-10	7.6E-10
U-238	4.47E+09 y	0.040	3.4E-07	0.020	1.2E-07	8.0E-08	6.8E-08	6.7E-08	4.5E-08
U-239	0.392 h	0.040	3.4E-10	0.020	1.9E-10	9.3E-11	5.4E-11	3.5E-11	2.7E-11
U-240	14.1 h	0.040	1.3E-08	0.020	8.1E-09	4.1E-09	2.4E-09	1.4E-09	1.1E-09
Neptunium[a]									
Np-232	0.245 h	0.005	8.7E-11	5.0E-04	5.1E-11	2.7E-11	1.7E-11	1.2E-11	9.7E-12
Np-233	0.603 h	0.005	2.1E-11	5.0E-04	1.3E-11	6.6E-12	4.0E-12	2.8E-12	2.2E-12
Np-234	4.40 d	0.005	6.2E-09	5.0E-04	4.4E-09	2.4E-09	1.6E-09	1.0E-09	8.1E-10
Np-235	1.08 y	0.005	7.1E-10	5.0E-04	4.1E-10	2.0E-10	1.2E-10	6.8E-11	5.3E-11
Np-236	1.15E+05 y	0.005	1.9E-07	5.0E-04	2.4E-08	1.8E-08	1.8E-08	1.8E-08	1.7E-08

a   Dose coefficients for this element are based on age-specific biokinetic data

Table A.1.—(continued)

Nuclide	Physical half-life	f_1 <1y	$e(\tau)$ 3 months	f_1 ≥1y	$e(\tau)$ 1 Year	5 Years	10 Years	15 Years	Adult
Np-236	22.5 h	0.005	2.5E-09	5.0E-04	1.3E-09	6.6E-10	4.0E-10	2.4E-10	1.9E-10
Np-237	2.14E+06 y	0.005	2.0E-06	5.0E-04	2.1E-07	1.4E-07	1.1E-07	1.1E-07	1.1E-07
Np-238	2.12 d	0.005	9.5E-09	5.0E-04	6.2E-09	3.2E-09	1.9E-09	1.1E-09	9.1E-10
Np-239	2.36 d	0.005	8.9E-09	5.0E-04	5.7E-09	2.9E-09	1.7E-09	1.0E-09	8.0E-10
Np-240	1.08 h	0.005	8.7E-10	5.0E-04	5.2E-10	2.6E-10	1.6E-10	1.0E-10	8.2E-11
Plutonium[*]									
Pu-234	8.80 h	0.005	2.1E-09	5.0E-04	1.1E-09	5.5E-10	3.3E-10	2.0E-10	1.6E-10
Pu-235	0.422 h	0.005	2.2E-11	5.0E-04	1.3E-11	6.5E-12	3.9E-12	2.7E-12	2.1E-12
Pu-236	2.85 y	0.005	2.1E-06	5.0E-04	2.2E-07	1.4E-07	1.0E-07	8.5E-08	8.7E-08
Pu-237	45.3 d	0.005	1.1E-09	5.0E-04	6.9E-10	3.6E-10	2.2E-10	1.3E-10	1.0E-10
Pu-238	87.7 y	0.005	4.0E-06	5.0E-04	4.0E-07	3.1E-07	2.4E-07	2.2E-07	2.3E-07
Pu-239	2.41E+04 y	0.005	4.2E-06	5.0E-04	4.2E-07	3.3E-07	2.7E-07	2.4E-07	2.5E-07
Pu-240	6.54E+03 y	0.005	4.2E-06	5.0E-04	4.2E-07	3.3E-07	2.7E-07	2.4E-07	2.5E-07
Pu-241	14.4 y	0.005	5.6E-08	5.0E-04	5.7E-09	5.5E-09	5.1E-09	4.8E-09	4.8E-09
Pu-242	3.76E+05 y	0.005	4.0E-06	5.0E-04	4.0E-07	3.2E-07	2.6E-07	2.3E-07	2.4E-07
Pu-243	4.95 h	0.005	1.0E-09	5.0E-04	6.2E-10	3.1E-10	1.8E-10	1.1E-10	8.5E-11
Pu-244	8.26E+07 y	0.005	4.0E-06	5.0E-04	4.1E-07	3.2E-07	2.6E-07	2.3E-07	2.4E-07
Pu-245	10.5 h	0.005	8.0E-09	5.0E-04	5.1E-09	2.6E-09	1.5E-09	8.9E-10	7.2E-10
Pu-246	10.9 d	0.005	3.6E-08	5.0E-04	2.3E-08	1.2E-08	7.1E-09	4.1E-09	3.3E-09
Americium[*]									
Am-237	1.22 h	0.005	1.7E-10	5.0E-04	1.0E-10	5.5E-11	3.3E-11	2.2E-11	1.8E-11
Am-238	1.63 h	0.005	2.5E-10	5.0E-04	1.6E-10	9.1E-11	5.9E-11	4.0E-11	3.2E-11
Am-239	11.9 h	0.005	2.6E-09	5.0E-04	1.7E-09	8.4E-10	5.1E-10	3.0E-10	2.4E-10
Am-240	2.12 d	0.005	4.7E-09	5.0E-04	3.3E-09	1.8E-09	1.2E-09	7.3E-10	5.8E-10
Am-241	4.32E+02 y	0.005	3.7E-06	5.0E-04	3.7E-07	2.7E-07	2.2E-07	2.0E-07	2.0E-07
Am-242	16.0 h	0.005	5.0E-09	5.0E-04	2.2E-09	1.1E-09	6.4E-10	3.7E-10	3.0E-10
Am-242m	1.52E+02 y	0.005	3.1E-06	5.0E-04	3.0E-07	2.3E-07	2.0E-07	1.9E-07	1.9E-07
Am-243	7.38E+03 y	0.005	3.6E-06	5.0E-04	3.7E-07	2.7E-07	2.2E-07	2.0E-07	2.0E-07
Am-244	10.1 h	0.005	4.9E-09	5.0E-04	3.1E-09	1.6E-09	9.6E-10	5.8E-10	4.6E-10
Am-244m	0.433 h	0.005	3.7E-10	5.0E-04	2.0E-10	9.6E-11	5.5E-11	3.7E-11	2.9E-11

[*] Dose coefficients for this element are based on age-specific biokinetic data

Table A.1.—*(continued)*

Nuclide	Physical half-life	f_1 <1y	$e(\tau)$ 3 months	f_1 ≥1y	$e(\tau)$ 1 Year	5 Years	10 Years	15 Years	Adult
Am-245	2.05 h	0.005	6.8E-10	5.0E-04	4.5E-10	2.2E-10	1.3E-10	7.9E-11	6.2E-11
Am-246	0.650 h	0.005	6.7E-10	5.0E-04	3.8E-10	1.9E-10	1.1E-10	7.3E-11	5.8E-11
Am-246m	0.417 h	0.005	3.9E-10	5.0E-04	2.2E-10	1.1E-10	6.4E-11	4.4E-11	3.4E-11
Curium[a]									
Cm-238	2.40 h	0.005	7.8E-10	5.0E-04	4.9E-10	2.6E-10	1.6E-10	1.0E-10	8.0E-11
Cm-240	27.0 d	0.005	2.2E-07	5.0E-04	4.8E-08	2.5E-08	1.5E-08	9.2E-09	7.6E-09
Cm-241	32.8 d	0.005	1.1E-08	5.0E-04	5.7E-09	3.0E-09	1.9E-09	1.1E-09	9.1E-10
Cm-242	163 d	0.005	5.9E-07	5.0E-04	7.6E-08	3.9E-08	2.4E-08	1.5E-08	1.2E-08
Cm-243	28.5 y	0.005	3.2E-06	5.0E-04	3.3E-07	2.2E-07	1.6E-07	1.4E-07	1.5E-07
Cm-244	18.1 y	0.005	2.9E-06	5.0E-04	2.9E-07	1.9E-07	1.4E-07	1.2E-07	1.2E-07
Cm-245	8.50E+03 y	0.005	3.7E-06	5.0E-04	3.7E-07	2.8E-07	2.3E-07	2.1E-07	2.1E-07
Cm-246	4.73E+03 y	0.005	3.7E-06	5.0E-04	3.7E-07	2.8E-07	2.2E-07	2.1E-07	2.1E-07
Cm-247	1.56E+07 y	0.005	3.4E-06	5.0E-04	3.5E-07	2.6E-07	2.1E-07	1.9E-07	1.9E-07
Cm-248	3.39E+05 y	0.005	1.4E-05	5.0E-04	1.4E-06	1.0E-06	8.4E-07	7.7E-07	7.7E-07
Cm-249	1.07 h	0.005	3.9E-10	5.0E-04	2.2E-10	1.1E-10	6.1E-11	4.0E-11	3.1E-11
Cm-250	6.90E+03 y	0.005	7.8E-05	5.0E-04	8.2E-06	6.0E-06	4.9E-06	4.4E-06	4.4E-06
Berkelium									
Bk-245	4.94 d	0.005	6.1E-09	5.0E-04	3.9E-09	2.0E-09	1.2E-09	7.2E-10	5.7E-10
Bk-246	1.83 d	0.005	3.7E-09	5.0E-04	2.6E-09	1.4E-09	9.4E-10	6.0E-10	4.8E-10
Bk-247	1.38E+03 y	0.005	8.9E-06	5.0E-04	8.6E-07	6.3E-07	4.6E-07	3.8E-07	3.5E-07
Bk-249	320 d	0.005	2.2E-08	5.0E-04	2.9E-09	1.9E-09	1.4E-09	1.1E-09	9.7E-10
Bk-250	3.22 h	0.005	1.5E-09	5.0E-04	8.5E-10	4.4E-10	2.7E-10	1.7E-10	1.4E-10
Californium									
Cf-244	0.323 h	0.005	9.8E-10	5.0E-04	4.8E-10	2.4E-10	1.3E-10	8.9E-11	7.0E-11
Cf-246	1.49 d	0.005	5.0E-08	5.0E-04	2.4E-08	1.2E-08	7.3E-09	4.1E-09	3.3E-09
Cf-248	334 d	0.005	1.5E-06	5.0E-04	1.6E-07	9.9E-08	6.0E-08	3.3E-08	2.8E-08
Cf-249	3.50E+02 y	0.005	9.0E-06	5.0E-04	8.7E-07	6.4E-07	4.7E-07	3.8E-07	3.5E-07
Cf-250	13.1 y	0.005	5.7E-06	5.0E-04	5.5E-07	3.7E-07	2.3E-07	1.7E-07	1.6E-07
Cf-251	8.98E+02 y	0.005	9.1E-06	5.0E-04	8.8E-07	6.5E-07	4.7E-07	3.9E-07	3.6E-07
Cf-252	2.64 y	0.005	5.0E-06	5.0E-04	5.1E-07	3.2E-07	1.9E-07	1.0E-07	9.0E-08

a   Dose coefficients for this element are based on **age-specific biokinetic data**

Table A.1.—(continued)

Nuclide	Physical half-life	f_1 <1y	$e(\tau)$ 3 months	f_1 ≥1y	$e(\tau)$ 1 Year	5 Years	10 Years	15 Years	Adult
Cf-253	17.8 d	0.005	1.0E-07	5.0E-04	1.1E-08	6.0E-09	3.7E-09	1.8E-09	1.4E-09
Cf-254	60.5 d	0.005	1.1E-05	5.0E-04	2.6E-06	1.4E-06	8.4E-07	5.0E-07	4.0E-07
Einsteinium									
Es-250	2.10 h	0.005	2.3E-10	5.0E-04	9.9E-11	5.7E-11	3.7E-11	2.6E-11	2.1E-11
Es-251	1.38 d	0.005	1.9E-09	5.0E-04	1.2E-09	6.1E-10	3.7E-10	2.2E-10	1.7E-10
Es-253	20.5 d	0.005	1.7E-07	5.0E-04	4.5E-08	2.3E-08	1.4E-08	7.6E-09	6.1E-09
Es-254	276 d	0.005	1.4E-06	5.0E-04	1.6E-07	9.8E-08	6.0E-08	3.3E-08	2.8E-08
Es-254m	1.64 d	0.005	5.7E-08	5.0E-04	3.0E-08	1.5E-08	9.1E-09	5.2E-09	4.2E-09
Fermium									
Fm-252	22.7 h	0.005	3.8E-08	5.0E-04	2.0E-08	9.9E-09	5.9E-09	3.3E-09	2.7E-09
Fm-253	3.00 d	0.005	2.5E-08	5.0E-04	6.7E-09	3.4E-09	2.1E-09	1.1E-09	9.1E-10
Fm-254	3.24 h	0.005	5.6E-09	5.0E-04	3.2E-09	1.6E-09	9.3E-10	5.6E-10	4.4E-10
Fm-255	20.1 h	0.005	3.3E-08	5.0E-04	1.9E-08	9.5E-09	5.6E-09	3.2E-09	2.5E-09
Fm-257	101 d	0.005	9.8E-07	5.0E-04	1.1E-07	6.5E-08	4.0E-08	1.9E-08	1.5E-08
Mendelevium									
Md-257	5.20 h	0.005	3.1E-09	5.0E-04	8.8E-10	4.5E-10	2.7E-10	1.5E-10	1.2E-10
Md-258	55.0 d	0.005	6.3E-07	5.0E-04	8.9E-08	5.0E-08	3.0E-08	1.6E-08	1.3E-08

Nuclide	Physical half-life	Type	f_1 <1y	$e(\tau)$ 3 Months	f_1 ≥1y	$e(\tau)$ 1 Year	5 Years	10 Years	15 Years	Adult
Hydrogen[a]										
Tritium compounds	12.3 y	F	1.000	2.6E-11	1.000	2.0E-11	1.1E-11	8.2E-12	5.9E-12	6.2E-12
		M	0.200	3.4E-10	0.100	2.7E-10	1.4E-10	8.2E-11	5.3E-11	4.5E-11
		S	0.020	1.2E-09	0.010	1.0E-09	6.3E-10	3.8E-10	2.8E-10	2.6E-10
Beryllium										
Be-7	53.3 d	M	0.020	2.5E-10	0.005	2.1E-10	1.2E-10	8.3E-11	6.2E-11	5.0E-11
		S	0.020	2.8E-10	0.005	2.4E-10	1.4E-10	9.6E-11	6.8E-11	5.5E-11
Be-10	1.60E+06 y	M	0.020	4.1E-08	0.005	3.4E-08	2.0E-08	1.3E-08	1.1E-08	9.6E-09
		S	0.020	9.9E-08	0.005	9.1E-08	6.1E-08	4.2E-08	3.7E-08	3.5E-08
Carbon[a]										
C-11	0.340 h	F	1.000	1.0E-10	1.000	7.0E-11	3.2E-11	2.1E-11	1.3E-11	1.1E-11
		M	0.200	1.5E-10	0.100	1.1E-10	4.9E-11	3.2E-11	2.1E-11	1.8E-11
		S	0.020	1.6E-10	0.010	1.1E-10	5.1E-11	3.3E-11	2.2E-11	1.8E-11
C-14	5.73E+03 y	F	1.000	6.1E-10	1.000	6.7E-10	3.6E-10	2.9E-10	1.9E-10	2.0E-10
		M	0.200	8.3E-09	0.100	6.6E-09	4.0E-09	2.8E-09	2.5E-09	2.0E-09
		S	0.020	1.9E-08	0.010	1.7E-08	1.1E-08	7.4E-09	6.4E-09	5.8E-09
Fluorine										
F-18	1.83 h	F	1.000	2.6E-10	1.000	1.9E-10	9.1E-11	5.6E-11	3.4E-11	2.8E-11
		M	1.000	4.1E-10	1.000	2.9E-10	1.5E-10	9.7E-11	6.9E-11	5.6E-11
		S	1.000	4.2E-10	1.000	3.1E-10	1.5E-10	1.0E-10	7.3E-11	5.9E-11
Sodium										
Na-22	2.60 y	F	1.000	9.7E-09	1.000	7.3E-09	3.8E-09	2.4E-09	1.5E-09	1.3E-09
Na-24	15.0 h	F	1.000	2.3E-09	1.000	1.8E-09	9.3E-10	5.7E-10	3.4E-10	2.7E-10
Magnesium										
Mg-28	20.9 h	F	1.000	5.3E-09	0.500	4.7E-09	2.2E-09	1.3E-09	7.3E-10	6.0E-10
		M	1.000	7.3E-09	0.500	7.2E-09	3.5E-09	2.3E-09	1.5E-09	1.2E-09
Aluminium										
Al-26	7.16E+05 y	F	0.020	8.1E-08	0.010	6.2E-08	3.2E-08	2.0E-08	1.3E-08	1.1E-08
		M	0.020	8.8E-08	0.010	7.4E-08	4.4E-08	2.9E-08	2.2E-08	2.0E-08
Silicon										
Si-31	2.62 h	F	0.020	3.6E-10	0.010	2.3E-10	9.5E-11	5.9E-11	3.2E-11	2.7E-11
		M	0.020	6.9E-10	0.010	4.4E-10	2.0E-10	1.3E-10	8.9E-11	7.4E-11
		S	0.020	7.2E-10	0.010	4.7E-10	2.2E-10	1.4E-10	9.5E-11	7.9E-11
Si-32	4.50E+02 y	F	0.020	3.0E-08	0.010	2.3E-08	1.1E-08	6.4E-09	3.8E-09	3.2E-09
		M	0.020	7.1E-08	0.010	6.0E-08	3.6E-08	2.4E-08	1.9E-08	1.7E-08
		S	0.020	2.8E-07	0.010	2.7E-07	1.9E-07	1.3E-07	1.1E-07	1.1E-07
Phosphorus										
P-32	14.3 d	F	1.000	1.2E-08	0.800	7.5E-09	3.2E-09	1.8E-09	9.8E-10	7.7E-10
		M	1.000	2.2E-08	0.800	1.5E-08	8.0E-09	5.3E-09	4.0E-09	3.4E-09
P-33	25.4 d	F	1.000	1.2E-09	0.800	7.8E-10	3.0E-10	2.0E-10	1.1E-10	9.2E-11
		M	1.000	6.1E-09	0.800	4.6E-09	2.8E-09	2.1E-09	1.9E-09	1.5E-09

a   Dose coefficients for radioisotopes of this element are based on age-specific biokinetic data

Table A.2.—(continued)

Nuclide	Physical half-life	Type	f_1 <1y	e(τ) 3 Months	f_1 ≥1y	e(τ) 1 Year	5 Years	10 Years	15 Years	Adult
Sulphur[a]										
S-35 (inorganic)	87.4 d	F	1.000	5.5E-10	0.800	3.9E-10	1.8E-10	1.1E-10	6.0E-11	5.1E-11
		M	0.200	5.9E-09	0.100	4.5E-09	2.8E-09	2.0E-09	1.8E-09	1.4E-09
		S	0.020	7.7E-09	0.010	6.0E-09	3.6E-09	2.6E-09	2.3E-09	1.9E-09
Chlorine										
Cl-36	3.01E+05 a	F	1.000	3.9E-09	1.000	2.6E-09	1.1E-09	7.1E-10	3.9E-10	2.3E-10
		M	1.000	3.1E-08	1.000	2.6E-08	1.5E-08	1.0E-08	8.8E-09	7.3E-09
Cl-38	0.620 h	F	1.000	2.9E-10	1.000	1.9E-10	8.4E-11	5.1E-11	3.0E-11	2.5E-11
		M	1.000	4.7E-10	1.000	3.0E-10	1.4E-10	8.5E-11	5.4E-11	4.5E-11
Cl-39	0.927 h	F	1.000	2.7E-10	1.000	1.8E-10	8.4E-11	5.1E-11	3.1E-11	2.5E-11
		M	1.000	4.3E-10	1.000	2.8E-10	1.3E-10	8.5E-11	5.5E-11	4.6E-11
Potassium										
K-40	1.28E+09 y	F	1.000	2.4E-08	1.000	1.7E-08	7.5E-09	4.5E-09	2.5E-09	1.1E-09
K-42	12.4 h	F	1.000	1.6E-09	1.000	1.0E-09	4.4E-10	2.6E-10	1.5E-10	1.2E-10
K-43	22.6 h	F	1.000	1.3E-09	1.000	9.7E-10	4.7E-10	2.9E-10	1.7E-10	1.4E-10
K-44	0.369 h	F	1.000	2.2E-10	1.000	1.4E-10	6.5E-11	4.0E-11	2.4E-11	2.0E-11
K-45	0.333 h	F	1.000	1.5E-10	1.000	1.0E-10	4.8E-11	3.0E-11	1.8E-11	1.5E-11
Calcium[a,b]										
Ca-41	1.40E+05 y	F	0.600	6.7E-10	0.300	3.8E-10	2.6E-10	3.3E-10	3.3E-10	1.7E-10
		M	0.200	4.2E-10	0.100	2.6E-10	1.7E-10	1.7E-10	1.6E-10	9.5E-11
		S	0.020	6.7E-10	0.010	6.0E-10	3.8E-10	2.4E-10	1.9E-10	1.8E-10
Ca-45	163 d	F	0.600	5.7E-09	0.300	3.0E-09	1.4E-09	1.0E-09	7.6E-10	4.6E-10
		M	0.200	1.2E-08	0.100	8.8E-09	5.3E-09	3.9E-09	3.5E-09	2.7E-09
		S	0.020	1.5E-08	0.010	1.2E-08	7.2E-09	5.1E-09	4.6E-09	3.7E-09
Ca-47	4.53 d	F	0.600	4.9E-09	0.300	3.6E-09	1.7E-09	1.1E-09	6.1E-10	5.3E-10
		M	0.200	1.0E-08	0.100	7.7E-09	4.2E-09	2.9E-09	2.4E-09	1.9E-09
		S	0.020	1.2E-08	0.010	8.5E-09	4.6E-09	3.3E-09	2.6E-09	2.1E-09
Scandium										
Sc-43	3.89 h	S	0.001	9.3E-10	1.0E-04	6.7E-10	3.3E-10	2.2E-10	1.4E-10	1.1E-10
Sc-44	3.93 h	S	0.001	1.6E-09	1.0E-04	1.2E-09	5.6E-10	3.6E-10	2.3E-10	1.8E-10
Sc-44m	2.44 d	S	0.001	1.1E-08	1.0E-04	8.4E-09	4.2E-09	2.8E-09	1.7E-09	1.4E-09
Sc-46	83.8 d	S	0.001	2.8E-08	1.0E-04	2.3E-08	1.4E-08	9.8E-09	8.4E-09	6.3E-09
Sc-47	3.35 d	S	0.001	4.0E-09	1.0E-04	2.8E-09	1.5E-09	1.1E-09	9.2E-10	7.3E-10
Sc-48	1.82 d	S	0.001	7.8E-09	1.0E-04	5.9E-09	3.1E-09	2.0E-09	1.4E-09	1.1E-09
Sc-49	0.956 h	S	0.001	3.9E-10	1.0E-04	2.4E-10	1.1E-10	7.1E-11	4.7E-11	4.0E-11
Titanium										
Ti-44	47.3 y	F	0.020	3.1E-07	0.010	2.6E-07	1.5E-07	9.6E-08	6.6E-08	6.1E-08
		M	0.020	1.7E-07	0.010	1.5E-07	9.2E-08	5.9E-08	4.6E-08	4.2E-08
		S	0.020	3.2E-07	0.010	3.1E-07	2.1E-07	1.5E-07	1.3E-07	1.2E-07

a   Dose coefficients for this element are based on age-specific biokinetic data

b   The f_1 value for 1 to 15 year olds for Type F is 0.4

Table A.2.—(continued)

Nuclide	Physical half-life	Type	f_1 <1y	$e(\tau)$ 3 Months	f_1 ≥1y	$e(\tau)$ 1 Year	5 Years	10 Years	15 Years	Adult
Ti-45	3.08 h	F	0.020	4.4E-10	0.010	3.2E-10	1.5E-10	9.1E-11	5.1E-11	4.2E-11
		M	0.020	7.4E-10	0.010	5.2E-10	2.5E-10	1.6E-10	1.1E-10	8.8E-11
		S	0.020	7.7E-10	0.010	5.5E-10	2.7E-10	1.7E-10	1.1E-10	9.3E-11
Vanadium										
V-47	0.543 h	F	0.020	1.8E-10	0.010	1.2E-10	5.6E-11	3.5E-11	2.1E-11	1.7E-11
		M	0.020	2.8E-10	0.010	1.9E-10	8.6E-11	5.5E-11	3.5E-11	2.9E-11
V-48	16.2 d	F	0.020	8.4E-09	0.010	6.4E-09	3.3E-09	2.1E-09	1.3E-09	1.1E-09
		M	0.020	1.4E-08	0.010	1.1E-08	6.3E-09	4.3E-09	2.9E-09	2.4E-09
V-49	330 d	F	0.020	2.0E-10	0.010	1.6E-10	7.7E-11	4.3E-11	2.5E-11	2.1E-11
		M	0.020	2.8E-10	0.010	2.1E-10	1.1E-10	6.3E-11	·4.0E-11	3.4E-11
Chromium										
Cr-48	23.0 h	F	0.200	7.6E-10	0.100	6.0E-10	3.1E-10	2.0E-10	1.2E-10	9.9E-11
		M	0.200	1.1E-09	0.100	9.1E-10	5.1E-10	3.4E-10	2.5E-10	2.0E-10
		S	0.200	1.2E-09	0.100	9.8E-10	5.5E-10	3.7E-10	2.8E-10	2.2E-10
Cr-49	0.702 h	F	0.200	1.9E-10	0.100	1.3E-10	6.0E-11	3.7E-11	2.2E-11	1.9E-11
		M	0.200	3.0E-10	0.100	2.0E-10	9.5E-11	6.1E-11	4.0E-11	3.3E-11
		S	0.200	3.1E-10	0.100	2.1E-10	9.9E-11	6.4E-11	4.2E-11	3.5E-11
Cr-51	27.7 d	F	0.200	1.7E-10	0.100	1.3E-10	6.3E-11	4.0E-11	2.4E-11	2.0E-11
		M	0.200	2.6E-10	0.100	1.9E-10	1.0E-10	6.4E-11	3.9E-11	3.2E-11
		S	0.200	2.6E-10	0.100	2.1E-10	1.0E-10	6.6E-11	4.5E-11	3.7E-11
Manganese										
Mn-51	0.770 h	F	0.200	2.5E-10	0.100	1.7E-10	7.5E-11	4.6E-11	2.7E-11	2.3E-11
		M	0.200	4.0E-10	0.100	2.7E-10	1.2E-10	7.8E-11	5.0E-11	4.1E-11
Mn-52	5.59 d	F	0.200	7.0E-09	0.100	5.5E-09	2.9E-09	1.8E-09	1.1E-09	9.4E-10
		M	0.200	8.6E-09	0.100	6.8E-09	3.7E-09	2.4E-09	1.7E-09	1.4E-09
Mn-52m	0.352 h	F	0.200	1.9E-10	0.100	1.3E-10	6.1E-11	3.8E-11	2.2E-11	1.9E-11
		M	0.200	2.8E-10	0.100	1.9E-10	8.7E-11	5.5E-11	3.4E-11	2.9E-11
Mn-53	3.70E+06 y	F	0.200	3.2E-10	0.100	2.2E-10	1.1E-10	6.0E-11	3.4E-11	2.9E-11
		M	0.200	4.6E-10	0.100	3.4E-10	1.7E-10	1.0E-10	6.4E-11	5.4E-11
Mn-54	312 d	F	0.200	5.2E-09	0.100	4.1E-09	2.2E-09	1.5E-09	9.9E-10	8.5E-10
		M	0.200	7.5E-09	0.100	6.2E-09	3.8E-09	2.4E-09	1.9E-09	1.5E-09
Mn-56	2.58 h	F	0.200	6.9E-10	0.100	4.9E-10	2.3E-10	1.4E-10	7.8E-11	6.4E-11
		M	0.200	1.1E-09	0.100	7.8E-10	3.7E-10	2.4E-10	1.5E-10	1.2E-10
Iron[a,b]										
Fe-52	8.28 h	F	0.600	5.2E-09	0.100	3.6E-09	1.5E-09	8.9E-10	4.9E-10	3.9E-10
		M	0.200	5.8E-09	0.100	4.1E-09	1.9E-09	1.2E-09	7.4E-10	6.0E-10
		S	0.020	6.0E-09	0.010	4.2E-09	2.0E-09	1.3E-09	7.7E-10	6.3E-10
Fe-55	2.70 y	F	0.600	4.2E-09	0.100	3.2E-09	2.2E-09	1.4E-09	9.4E-10	7.7E-10
		M	0.200	1.9E-09	0.100	1.4E-09	9.9E-10	6.2E-10	4.4E-10	3.8E-10
		S	0.020	1.0E-09	0.010	8.5E-10	5.0E-10	2.9E-10	2.0E-10	1.8E-10
Fe-59	44.5 d	F	0.600	2.1E-08	0.100	1.3E-08	7.1E-09	4.2E-09	2.6E-09	2.2E-09
		M	0.200	1.8E-08	0.100	1.3E-08	7.9E-09	5.5E-09	4.6E-09	3.7E-09
		S	0.020	1.7E-08	0.010	1.3E-08	8.1E-09	5.8E-09	5.1E-09	4.0E-09
Fe-60	1.00E+05 y	F	0.600	4.4E-07	0.100	3.9E-07	3.5E-07	3.2E-07	2.9E-07	2.8E-07
		M	0.200	2.0E-07	0.100	1.7E-07	1.6E-07	1.4E-07	1.4E-07	1.4E-07
		S	0.020	9.3E-08	0.010	8.8E-08	6.7E-08	5.2E-08	4.9E-08	4.9E-08

a Dose coefficients for this element are based on age-specific biokinetic data

b The f_1 value for 1 to 15 year olds for Type F is 0.2

Table A.2.—(continued)

Nuclide	Physical half-life	Type	f_1 <1y	$e(\tau)$ 3 Months	f_1 ≥1y	$e(\tau)$ 1 Year	5 Years	10 Years	15 Years	Adult
Cobalt[a,b]										
Co-55	17.5 h	F	0.600	2.2E-09	0.100	1.8E-09	9.0E-10	5.5E-10	3.1E-10	2.7E-10
		M	0.200	4.1E-09	0.100	3.1E-09	1.5E-09	9.8E-10	6.1E-10	5.0E-10
		S	0.020	4.6E-09	0.010	3.3E-09	1.6E-09	1.1E-09	6.6E-10	5.3E-10
Co-56	78.7 d	F	0.600	1.4E-08	0.100	1.0E-08	5.5E-09	3.5E-09	2.2E-09	1.8E-09
		M	0.200	2.5E-08	0.100	2.1E-08	1.1E-08	7.4E-09	5.8E-09	4.8E-09
		S	0.020	2.9E-08	0.010	2.5E-08	1.5E-08	1.0E-08	8.0E-09	6.7E-09
Co-57	271 d	F	0.600	1.5E-09	0.100	1.1E-09	5.6E-10	3.7E-10	2.3E-10	1.9E-10
		M	0.200	2.8E-09	0.100	2.2E-09	1.3E-09	8.5E-10	6.7E-10	5.5E-10
		S	0.020	4.4E-09	0.010	3.7E-09	2.3E-09	1.5E-09	1.2E-09	1.0E-09
Co-58	70.8 d	F	0.600	4.0E-09	0.100	3.0E-09	1.6E-09	1.0E-09	6.4E-10	5.3E-10
		M	0.200	7.3E-09	0.100	6.5E-09	3.5E-09	2.4E-09	2.0E-09	1.6E-09
		S	0.020	9.0E-09	0.010	7.5E-09	4.5E-09	3.1E-09	2.6E-09	2.1E-09
Co-58m	9.15 h	F	0.600	4.8E-11	0.100	3.6E-11	1.7E-11	1.1E-11	5.9E-12	5.2E-12
		M	0.200	1.1E-10	0.100	7.6E-11	3.8E-11	2.4E-11	1.6E-11	1.3E-11
		S	0.020	1.3E-10	0.010	9.0E-11	4.5E-11	3.0E-11	2.0E-11	1.7E-11
Co-60	5.27 y	F	0.600	3.0E-08	0.100	2.3E-08	1.4E-08	8.9E-09	6.1E-09	5.2E-09
		M	0.200	4.2E-08	0.100	3.4E-08	2.1E-08	1.5E-08	1.2E-08	1.0E-08
		S	0.020	9.2E-08	0.010	8.6E-08	5.9E-08	4.0E-08	3.4E-08	3.1E-08
Co-60m	0.174 h	F	0.600	4.4E-12	0.100	2.8E-12	1.5E-12	1.0E-12	8.3E-13	6.9E-13
		M	0.200	7.1E-12	0.100	4.7E-12	2.7E-12	1.8E-12	1.5E-12	1.2E-12
		S	0.020	7.6E-12	0.010	5.1E-12	2.9E-12	2.0E-12	1.7E-12	1.4E-12
Co-61	1.65 h	F	0.600	2.1E-10	0.100	1.4E-10	6.0E-11	3.8E-11	2.2E-11	1.9E-11
		M	0.200	4.0E-10	0.100	2.7E-10	1.2E-10	8.2E-11	5.7E-11	4.7E-11
		S	0.020	4.3E-10	0.010	2.8E-10	1.3E-10	8.8E-11	6.1E-11	5.1E-11
Co-62m	0.232 h	F	0.600	1.4E-10	0.100	9.5E-11	4.5E-11	2.8E-11	1.7E-11	1.4E-11
		M	0.200	1.9E-10	0.100	1.3E-10	6.1E-11	3.8E-11	2.4E-11	2.0E-11
		S	0.020	2.0E-10	0.010	1.3E-10	6.3E-11	4.0E-11	2.5E-11	2.1E-11
Nickel[a]										
Ni-56	6.10 d	F	0.100	3.3E-09	0.050	2.8E-09	1.5E-09	9.3E-10	5.8E-10	4.9E-10
		M	0.100	4.9E-09	0.050	4.1E-09	2.3E-09	1.5E-09	1.1E-09	8.7E-10
		S	0.020	5.5E-09	0.010	4.6E-09	2.7E-09	1.8E-09	1.3E-09	1.0E-09
Ni-57	1.50 d	F	0.100	2.2E-09	0.050	1.8E-09	8.9E-10	5.5E-10	3.1E-10	2.5E-10
		M	0.100	3.6E-09	0.050	2.8E-09	1.5E-09	9.5E-10	6.2E-10	5.0E-10
		S	0.020	3.9E-09	0.010	3.0E-09	1.5E-09	1.0E-09	6.6E-10	5.3E-10
Ni-59	7.50E+04 y	F	0.100	9.6E-10	0.050	8.1E-10	4.5E-10	2.8E-10	1.9E-10	1.9E-10
		M	0.100	7.9E-10	0.050	6.2E-10	3.4E-10	2.1E-10	1.4E-10	1.3E-10
		S	0.020	1.7E-09	0.010	1.5E-09	9.5E-10	5.9E-10	4.6E-10	4.4E-10
Ni-63	96.0 y	F	0.100	2.3E-09	0.050	2.0E-09	1.1E-09	6.7E-10	4.6E-10	4.4E-10
		M	0.100	2.5E-09	0.050	1.9E-09	1.1E-09	7.0E-10	5.3E-10	4.8E-10
		S	0.020	4.8E-09	0.010	4.3E-09	2.7E-09	1.7E-09	1.3E-09	1.3E-09
Ni-65	2.52 h	F	0.100	4.4E-10	0.050	3.0E-10	1.4E-10	8.5E-11	4.9E-11	4.1E-11
		M	0.100	7.7E-10	0.050	5.2E-10	2.4E-10	1.6E-10	1.0E-10	8.5E-11
		S	0.020	8.1E-10	0.010	5.5E-10	2.6E-10	1.7E-10	1.1E-10	9.0E-11
Ni-66	2.27 d	F	0.100	5.7E-09	0.050	3.8E-09	1.6E-09	1.0E-09	5.1E-10	4.2E-10
		M	0.100	1.3E-08	0.050	9.4E-09	4.5E-09	2.9E-09	2.0E-09	1.6E-09
		S	0.020	1.5E-08	0.010	1.0E-08	5.0E-09	3.2E-09	2.2E-09	1.8E-09

a Dose coefficients for this element are based on age-specific biokinetic data

b The f_1 value for 1 to 15 year olds for Type F is 0.3

Table A.2.—(continued)

Nuclide	Physical half-life	Type	f_1 <1y	$e(\tau)$ 3 Months	f_1 ≥1y	$e(\tau)$ 1 Year	5 Years	10 Years	15 Years	Adult
Copper										
Cu-60	0.387 h	F	1.000	2.1E-10	0.500	1.6E-10	7.5E-11	4.6E-11	2.8E-11	2.3E-11
		M	1.000	3.0E-10	0.500	2.2E-10	1.0E-10	6.5E-11	4.0E-11	3.3E-11
		S	1.000	3.1E-10	0.500	2.2E-10	1.1E-10	6.7E-11	4.2E-11	3.4E-11
Cu-61	3.41 h	F	1.000	3.1E-10	0.500	2.7E-10	1.3E-10	7.9E-11	4.5E-11	3.7E-11
		M	1.000	4.9E-10	0.500	4.4E-10	2.1E-10	1.4E-10	9.1E-11	7.4E-11
		S	1.000	5.1E-10	0.500	4.5E-10	2.2E-10	1.4E-10	9.6E-11	7.8E-11
Cu-64	12.7 h	F	1.000	2.8E-10	0.500	2.7E-10	1.2E-10	7.6E-11	4.2E-11	3.5E-11
		M	1.000	5.5E-10	0.500	5.4E-10	2.7E-10	1.9E-10	1.4E-10	1.1E-10
		S	1.000	5.8E-10	0.500	5.7E-10	2.9E-10	2.0E-10	1.3E-10	1.2E-10
Cu-67	2.58 d	F	1.000	9.5E-10	0.500	8.0E-10	3.5E-10	2.2E-10	1.2E-10	1.0E-10
		M	1.000	2.3E-09	0.500	2.0E-09	1.1E-09	8.1E-10	6.9E-10	5.5E-10
		S	1.000	2.5E-09	0.500	2.1E-09	1.2E-09	8.9E-10	7.7E-10	6.1E-10
Zinc[a]										
Zn-62	9.26 h	F	1.000	1.7E-09	0.500	1.7E-09	7.7E-10	4.6E-10	2.5E-10	2.0E-10
		M	0.200	4.5E-09	0.100	3.5E-09	1.6E-09	1.0E-09	6.0E-10	5.0E-10
		S	0.020	5.1E-09	0.010	3.4E-09	1.8E-09	1.1E-09	6.6E-10	5.5E-10
Zn-63	0.635 h	F	1.000	2.1E-10	0.500	1.4E-10	6.5E-11	4.0E-11	2.4E-11	2.0E-11
		M	0.200	3.4E-10	0.100	2.3E-10	1.0E-10	6.6E-11	4.2E-11	3.5E-11
		S	0.020	3.6E-10	0.010	2.4E-10	1.1E-10	6.9E-11	4.4E-11	3.7E-11
Zn-65	244 d	F	1.000	1.5E-08	0.500	1.0E-08	5.7E-09	3.8E-09	2.5E-09	2.2E-09
		M	0.200	8.5E-09	0.100	6.5E-09	3.7E-09	2.4E-09	1.9E-09	1.6E-09
		S	0.020	7.6E-09	0.010	6.7E-09	4.4E-09	2.9E-09	2.4E-09	2.0E-09
Zn-69	0.950 h	F	1.000	1.1E-10	0.500	7.4E-11	3.2E-11	2.1E-11	1.2E-11	1.1E-11
		M	0.200	2.2E-10	0.100	1.4E-10	6.5E-11	4.4E-11	3.1E-11	2.6E-11
		S	0.020	2.3E-10	0.010	1.5E-10	6.9E-11	4.7E-11	3.4E-11	2.8E-11
Zn-69m	13.8 h	F	1.000	6.6E-10	0.500	6.7E-10	3.0E-10	1.8E-10	9.9E-11	8.2E-11
		M	0.200	2.1E-09	0.100	1.5E-09	7.5E-10	5.0E-10	3.0E-10	2.4E-10
		S	0.020	2.2E-09	0.010	1.7E-09	8.2E-10	5.4E-10	3.3E-10	2.7E-10
Zn-71m	3.92 h	F	1.000	6.2E-10	0.500	5.5E-10	2.6E-10	1.6E-10	9.1E-11	7.4E-11
		M	0.200	1.3E-09	0.100	9.4E-10	4.6E-10	2.9E-10	1.9E-10	1.5E-10
		S	0.020	1.4E-09	0.010	1.0E-09	4.9E-10	3.1E-10	2.0E-10	1.6E-10
Zn-72	1.94 d	F	1.000	4.3E-09	0.500	3.5E-09	1.7E-09	1.0E-09	5.9E-10	4.9E-10
		M	0.200	8.8E-09	0.100	6.5E-09	3.4E-09	2.3E-09	1.5E-09	1.2E-09
		S	0.020	9.7E-09	0.010	7.0E-09	3.6E-09	2.4E-09	1.6E-09	1.3E-09
Gallium										
Ga-65	0.253 h	F	0.010	1.1E-10	0.001	7.3E-11	3.4E-11	2.1E-11	1.3E-11	1.1E-11
		M	0.010	1.6E-10	0.001	1.1E-10	4.8E-11	3.1E-11	2.0E-11	1.7E-11
Ga-66	9.40 h	F	0.010	2.8E-09	0.001	2.0E-09	9.2E-10	5.7E-10	3.0E-10	2.5E-10
		M	0.010	4.5E-09	0.001	3.1E-09	1.5E-09	9.2E-10	5.3E-10	4.4E-10
Ga-67	3.26 d	F	0.010	6.4E-10	0.001	4.6E-10	2.2E-10	1.4E-10	7.7E-11	6.4E-11
		M	0.010	1.4E-09	0.001	1.0E-09	5.0E-10	3.6E-10	3.0E-10	2.4E-10
Ga-68	1.13 h	F	0.010	2.9E-10	0.001	1.9E-10	8.8E-11	5.4E-11	3.1E-11	2.6E-11
		M	0.010	4.6E-10	0.001	3.1E-10	1.4E-10	9.2E-11	5.9E-11	4.9E-11
Ga-70	0.353 h	F	0.010	9.5E-11	0.001	6.0E-11	2.6E-11	1.6E-11	1.0E-11	8.8E-12
		M	0.010	1.5E-10	0.001	9.6E-11	4.3E-11	2.8E-11	1.8E-11	1.6E-11
Ga-72	14.1 h	F	0.010	2.9E-09	0.001	2.2E-09	1.0E-09	6.4E-10	3.6E-10	2.9E-10
		M	0.010	4.5E-09	0.001	3.3E-09	1.6E-09	1.0E-09	6.5E-10	5.3E-10

a Dose coefficients for this element are based on age-specific biokinetic data

Table A.2.—(continued)

Nuclide	Physical half-life	Type	f_1 <1y	$e(\tau)$ 3 Months	f_1 ≥1y	$e(\tau)$ 1 Year	5 Years	10 Years	15 Years	Adult
Ca-73	4.91 h	F	0.010	6.7E-10	0.001	4.5E-10	2.0E-10	1.2E-10	6.4E-11	5.4E-11
		M	0.010	1.2E-09	0.001	8.4E-10	4.0E-10	2.6E-10	1.7E-10	1.4E-10
Germanium										
Ge-66	2.27 h	F	1.000	4.5E-10	1.000	3.5E-10	1.8E-10	1.1E-10	6.7E-11	5.4E-11
		M	1.000	6.4E-10	1.000	4.8E-10	2.5E-10	1.6E-10	1.1E-10	9.1E-11
Ge-67	0.312 h	F	1.000	1.7E-10	1.000	1.1E-10	4.9E-11	3.1E-11	1.8E-11	1.5E-11
		M	1.000	2.5E-10	1.000	1.6E-10	7.3E-11	4.6E-11	2.9E-11	2.5E-11
Ge-68	288 d	F	1.000	5.4E-09	1.000	3.8E-09	1.8E-09	1.1E-09	6.3E-10	5.2E-10
		M	1.000	6.0E-08	1.000	5.0E-08	3.0E-08	2.0E-08	1.6E-08	1.4E-08
Ge-69	1.63 d	F	1.000	1.2E-09	1.000	9.0E-10	4.6E-10	2.8E-10	1.7E-10	1.3E-10
		M	1.000	1.8E-09	1.000	1.4E-09	7.4E-10	4.9E-10	3.6E-10	2.9E-10
Ge-71	11.8 d	F	1.000	6.0E-11	1.000	4.3E-11	2.0E-11	1.1E-11	6.1E-12	4.8E-12
		M	1.000	1.2E-10	1.000	8.6E-11	4.1E-11	2.4E-11	1.3E-11	1.1E-11
Ge-75	1.38 h	F	1.000	1.6E-10	1.000	1.0E-10	4.3E-11	2.8E-11	1.7E-11	1.5E-11
		M	1.000	2.9E-10	1.000	1.9E-10	8.9E-11	6.1E-11	4.4E-11	3.6E-11
Ge-77	11.3 h	F	1.000	1.3E-09	1.000	9.5E-10	4.7E-10	2.9E-10	1.7E-10	1.4E-10
		M	1.000	2.3E-09	1.000	1.7E-09	8.8E-10	6.0E-10	4.5E-10	3.7E-10
Ge-78	1.45 h	F	1.000	4.3E-10	1.000	2.9E-10	1.4E-10	8.9E-11	5.5E-11	4.5E-11
		M	1.000	7.3E-10	1.000	5.0E-10	2.5E-10	1.6E-10	1.2E-10	9.5E-11
Arsenic										
As-69	0.253 h	M	1.000	2.1E-10	0.500	1.4E-10	6.3E-11	4.0E-11	2.5E-11	2.1E-11
As-70	0.876 h	M	1.000	5.7E-10	0.500	4.3E-10	2.1E-10	1.3E-10	8.3E-11	6.7E-11
As-71	2.70 d	M	1.000	2.2E-09	0.500	1.9E-09	1.0E-09	6.8E-10	5.0E-10	4.0E-10
As-72	1.08 d	M	1.000	5.9E-09	0.500	5.7E-09	2.7E-09	1.7E-09	1.1E-09	9.0E-10
As-73	80.3 d	M	1.000	5.4E-09	0.500	4.0E-09	2.3E-09	1.5E-09	1.2E-09	1.0E-09
As-74	17.8 d	M	1.000	1.1E-08	0.500	8.4E-09	4.7E-09	3.3E-09	2.6E-09	2.1E-09
As-76	1.10 d	M	1.000	5.1E-09	0.500	4.6E-09	2.2E-09	1.4E-09	8.8E-10	7.4E-10
As-77	1.62 d	M	1.000	2.2E-09	0.500	1.7E-09	8.9E-10	6.2E-10	5.0E-10	4.9E-10
As-78	1.51 h	M	1.000	8.0E-10	0.500	5.8E-10	2.7E-10	1.7E-10	1.1E-10	8.9E-11
Selenium[a]										
Se-70	0.683 h	F	1.000	3.9E-10	0.800	3.0E-10	1.5E-10	9.0E-11	5.1E-11	4.2E-11
		M	0.200	6.5E-10	0.100	4.7E-10	2.3E-10	1.4E-10	8.9E-11	7.3E-11
		S	0.020	6.8E-10	0.010	4.8E-10	2.3E-10	1.5E-10	9.4E-11	7.6E-11
Se-73	7.15 h	F	1.000	7.7E-10	0.800	6.5E-10	3.3E-10	2.1E-10	1.0E-10	8.0E-11
		M	0.200	1.6E-09	0.100	1.2E-09	5.9E-10	3.8E-10	2.4E-10	1.9E-10
		S	0.020	1.8E-09	0.010	1.3E-09	6.3E-10	4.0E-10	2.6E-10	2.1E-10
Se-73m	0.650 h	F	1.000	9.3E-11	0.800	7.2E-11	3.5E-11	2.3E-11	1.1E-11	9.2E-12
		M	0.200	1.8E-10	0.100	1.3E-10	6.1E-11	3.9E-11	2.5E-11	2.0E-11
		S	0.020	1.9E-10	0.010	1.3E-10	6.5E-11	4.1E-11	2.6E-11	2.2E-11
Se-75	120 d	F	1.000	7.8E-09	0.800	6.0E-09	3.4E-09	2.5E-09	1.2E-09	1.0E-09
		M	0.200	5.4E-09	0.100	4.5E-09	2.5E-09	1.7E-09	1.3E-09	1.1E-09
		S	0.020	5.6E-09	0.010	4.7E-09	2.9E-09	2.0E-09	1.6E-09	1.3E-09

a Dose coefficients for this element are based on age-specific biokinetic data

Table A.2.—(continued)

Nuclide	Physical half-life	Type	f_1 <1y	$e(\tau)$ 3 Months	f_1 ≥1y	$e(\tau)$ 1 Year	5 Years	10 Years	15 Years	Adult
Se-79	6.50E+04 y	F	1.000	1.6E-08	0.800	1.3E-08	7.7E-09	5.6E-09	1.5E-09	1.1E-09
		M	0.200	1.4E-08	0.100	1.1E-08	6.9E-09	4.9E-09	3.3E-09	2.6E-09
		S	0.020	2.3E-08	0.010	2.0E-08	1.3E-08	8.7E-09	7.6E-09	6.8E-09
Se-81	0.308 h	F	1.000	8.6E-11	0.800	5.4E-11	2.3E-11	1.5E-11	9.2E-12	8.0E-12
		M	0.200	1.3E-10	0.100	8.5E-11	3.8E-11	2.5E-11	1.6E-11	1.4E-11
		S	0.020	1.4E-10	0.010	8.9E-11	3.9E-11	2.6E-11	1.7E-11	1.5E-11
Se-81m	0.954 h	F	1.000	1.8E-10	0.800	1.2E-10	5.4E-11	3.4E-11	1.9E-11	1.6E-11
		M	0.200	3.8E-10	0.100	2.5E-10	1.2E-10	8.0E-11	5.8E-11	4.7E-11
		S	0.020	4.1E-10	0.010	2.7E-10	1.3E-10	8.5E-11	6.2E-11	5.1E-11
Se-83	0.375 h	F	1.000	1.7E-10	0.800	1.2E-10	5.8E-11	3.6E-11	2.1E-11	1.8E-11
		M	0.200	2.7E-10	0.100	1.9E-10	9.2E-11	5.9E-11	3.9E-11	3.2E-11
		S	0.020	2.8E-10	0.010	2.0E-10	9.6E-11	6.2E-11	4.1E-11	3.4E-11
Bromine										
Br-74	0.422 h	F	1.000	2.5E-10	1.000	1.8E-10	8.6E-11	5.3E-11	3.2E-11	2.6E-11
		M	1.000	3.6E-10	1.000	2.5E-10	1.2E-10	7.5E-11	4.6E-11	3.8E-11
Br-74m	0.691 h	F	1.000	4.0E-10	1.000	2.8E-10	1.3E-10	8.1E-11	4.8E-11	3.9E-11
		M	1.000	5.9E-10	1.000	4.1E-10	1.9E-10	1.2E-10	7.5E-11	6.2E-11
Br-75	1.63 h	F	1.000	2.9E-10	1.000	2.1E-10	9.7E-11	5.9E-11	3.5E-11	2.9E-11
		M	1.000	4.5E-10	1.000	3.1E-10	1.5E-10	9.7E-11	6.5E-11	5.3E-11
Br-76	16.2 h	F	1.000	2.2E-09	1.000	1.7E-09	8.4E-10	5.1E-10	3.0E-10	2.4E-10
		M	1.000	3.0E-09	1.000	2.3E-09	1.2E-09	7.5E-10	5.0E-10	4.1E-10
Br-77	2.33 d	F	1.000	5.3E-10	1.000	4.4E-10	2.2E-10	1.3E-10	7.7E-11	6.2E-11
		M	1.000	6.3E-10	1.000	5.1E-10	2.7E-10	1.6E-10	1.1E-10	8.4E-11
Br-80	0.290 h	F	1.000	7.1E-11	1.000	4.4E-11	1.8E-11	1.2E-11	6.9E-12	5.9E-12
		M	1.000	1.1E-10	1.000	6.5E-11	2.8E-11	1.8E-11	1.1E-11	9.4E-12
Br-80m	4.42 h	F	1.000	4.3E-10	1.000	2.8E-10	1.2E-10	7.2E-11	4.0E-11	3.3E-11
		M	1.000	6.8E-10	1.000	4.5E-10	2.1E-10	1.4E-10	9.3E-11	7.6E-11
Br-82	1.47 d	F	1.000	2.7E-09	1.000	2.2E-09	1.2E-09	7.0E-10	4.2E-10	3.5E-10
		M	1.000	3.8E-09	1.000	3.0E-09	1.7E-09	1.1E-09	7.9E-10	6.3E-10
Br-83	2.39 h	F	1.000	1.7E-10	1.000	1.1E-10	4.7E-11	3.0E-11	1.8E-11	1.6E-11
		M	1.000	3.5E-10	1.000	2.3E-10	1.1E-10	7.7E-11	5.9E-11	4.8E-11
Br-84	0.530 h	F	1.000	2.4E-10	1.000	1.6E-10	7.1E-11	4.4E-11	2.6E-11	2.2E-11
		M	1.000	3.7E-10	1.000	2.4E-10	1.1E-10	6.9E-11	4.4E-11	3.7E-11
Rubidium										
Rb-79	0.382 h	F	1.000	1.6E-10	1.000	1.1E-10	5.0E-11	3.2E-11	1.9E-11	1.6E-11
Rb-81	4.58 h	F	1.000	3.2E-10	1.000	2.5E-10	1.2E-10	7.1E-11	4.2E-11	3.4E-11
Rb-81m	0.533 h	F	1.000	6.2E-11	1.000	4.6E-11	2.2E-11	1.4E-11	8.5E-12	7.0E-12
Rb-82m	6.20 h	F	1.000	8.6E-10	1.000	7.3E-10	3.9E-10	2.3E-10	1.4E-10	1.1E-10
Rb-83	86.2 d	F	1.000	4.9E-09	1.000	3.8E-09	2.0E-09	1.3E-09	7.9E-10	6.9E-10
Rb-84	32.8 d	F	1.000	8.6E-09	1.000	6.4E-09	3.1E-09	2.0E-09	1.2E-09	1.0E-09
Rb-86	18.7 d	F	1.000	1.2E-08	1.000	7.7E-09	3.4E-09	2.0E-09	1.1E-09	9.3E-10
Rb-87	4.70E+10 y	F	1.000	6.0E-09	1.000	4.1E-09	1.8E-09	1.1E-09	6.0E-10	5.0E-10
Rb-88	0.297 h	F	1.000	1.9E-10	1.000	1.2E-10	5.2E-11	3.2E-11	1.9E-11	1.6E-11

Table A.2.—(continued)

Nuclide	Physical half-life	Type	f_1 <1y	$e(\tau)$ 3 Months	f_1 ≥1y	$e(\tau)$ 1 Year	5 Years	10 Years	15 Years	Adult
Rb-89	0.253 h	F	1.000	1.4E-10	1.000	9.3E-11	4.3E-11	2.7E-11	1.6E-11	1.4E-11
Strontium[a,b]										
Sr-80	1.67 h	F	0.600	7.8E-10	0.300	5.4E-10	2.4E-10	1.4E-10	7.9E-11	7.1E-11
		M	0.200	1.4E-09	0.100	9.0E-10	4.1E-10	2.5E-10	1.5E-10	1.3E-10
		S	0.020	1.5E-09	0.010	9.4E-10	4.3E-10	2.7E-10	1.6E-10	1.4E-10
Sr-81	0.425 h	F	0.600	2.1E-10	0.300	1.5E-10	6.7E-11	4.1E-11	2.4E-11	2.1E-11
		M	0.200	3.3E-10	0.100	2.2E-10	1.0E-10	6.6E-11	4.2E-11	3.5E-11
		S	0.020	3.4E-10	0.010	2.3E-10	1.1E-10	6.9E-11	4.4E-11	3.7E-11
Sr-82	25.0 d	F	0.600	2.8E-08	0.300	1.5E-08	6.6E-09	4.6E-09	3.2E-09	2.1E-09
		M	0.200	5.5E-08	0.100	4.0E-08	2.1E-08	1.4E-08	1.0E-08	8.9E-09
		S	0.020	6.1E-08	0.010	4.6E-08	2.5E-08	1.7E-08	1.2E-08	1.1E-08
Sr-83	1.35 d	F	0.600	1.4E-09	0.300	1.1E-09	5.5E-10	3.4E-10	2.0E-10	1.6E-10
		M	0.200	2.5E-09	0.100	1.9E-09	9.5E-10	6.0E-10	3.9E-10	3.1E-10
		S	0.020	2.8E-09	0.010	2.0E-09	1.0E-09	6.5E-10	4.2E-10	3.4E-10
Sr-85	64.8 d	F	0.600	4.4E-09	0.300	2.3E-09	1.1E-09	9.6E-10	8.3E-10	3.8E-10
		M	0.200	4.3E-09	0.100	3.1E-09	1.8E-09	1.2E-09	8.8E-10	6.4E-10
		S	0.020	4.4E-09	0.010	3.7E-09	2.2E-09	1.3E-09	1.0E-09	8.1E-10
Sr-85m	1.16 h	F	0.600	2.4E-11	0.300	1.9E-11	9.6E-12	6.0E-12	3.7E-12	2.9E-12
		M	0.200	3.1E-11	0.100	2.5E-11	1.3E-11	8.0E-12	5.1E-12	4.1E-12
		S	0.020	3.2E-11	0.010	2.6E-11	1.3E-11	8.3E-12	5.4E-12	4.3E-12
Sr-87m	2.80 h	F	0.600	9.7E-11	0.300	7.8E-11	3.8E-11	2.3E-11	1.3E-11	1.1E-11
		M	0.200	1.6E-10	0.100	1.2E-10	5.9E-11	3.8E-11	2.5E-11	2.0E-11
		S	0.020	1.7E-10	0.010	1.2E-10	6.2E-11	4.0E-11	2.6E-11	2.1E-11
Sr-89	50.5 d	F	0.600	1.5E-08	0.300	7.3E-09	3.2E-09	2.3E-09	1.7E-09	1.0E-09
		M	0.200	3.3E-08	0.100	2.4E-08	1.3E-08	9.1E-09	7.3E-09	6.1E-09
		S	0.020	3.9E-08	0.010	3.0E-08	1.7E-08	1.2E-08	9.3E-09	7.9E-09
Sr-90	29.1 y	F	0.600	1.3E-07	0.300	5.2E-08	3.1E-08	4.1E-08	5.3E-08	2.4E-08
		M	0.200	1.5E-07	0.100	1.1E-07	6.5E-08	5.1E-08	5.0E-08	3.6E-08
		S	0.020	4.2E-07	0.010	4.0E-07	2.7E-07	1.8E-07	1.6E-07	1.6E-07
Sr-91	9.50 h	F	0.600	1.4E-09	0.300	1.1E-09	5.2E-10	3.1E-10	1.7E-10	1.6E-10
		M	0.200	3.1E-09	0.100	2.2E-09	1.1E-09	6.9E-10	4.4E-10	3.7E-10
		S	0.020	3.5E-09	0.010	2.5E-09	1.2E-09	7.7E-10	4.9E-10	4.1E-10
Sr-92	2.71 h	F	0.600	9.0E-10	0.300	7.1E-10	3.3E-10	2.0E-10	1.0E-10	9.8E-11
		M	0.200	1.9E-09	0.100	1.4E-09	6.5E-10	4.1E-10	2.5E-10	2.1E-10
		S	0.020	2.2E-09	0.010	1.5E-09	7.0E-10	4.5E-10	2.7E-10	2.3E-10
Yttrium										
Y-86	14.7 h	M	0.001	3.7E-09	1.0E-04	2.9E-09	1.5E-09	9.3E-10	5.6E-10	4.5E-10
		S	0.001	3.8E-09	1.0E-04	3.0E-09	1.5E-09	9.6E-10	5.8E-10	4.7E-10
Y-86m	0.800 h	M	0.001	2.2E-10	1.0E-04	1.7E-10	8.7E-11	5.6E-11	3.4E-11	2.7E-11
		S	0.001	2.3E-10	1.0E-04	1.8E-10	9.0E-11	5.7E-11	3.5E-11	2.8E-11
Y-87	3.35 d	M	0.001	2.7E-09	1.0E-04	2.1E-09	1.1E-09	7.0E-10	4.7E-10	3.7E-10
		S	0.001	2.8E-09	1.0E-04	2.2E-09	1.1E-09	7.3E-10	5.0E-10	3.9E-10
Y-88	107 d	M	0.001	1.9E-08	1.0E-04	1.6E-08	1.0E-08	6.7E-09	4.9E-09	4.1E-09
		S	0.001	2.0E-08	1.0E-04	1.7E-08	9.8E-09	6.6E-09	5.4E-09	4.4E-09
Y-90	2.67 d	M	0.001	1.3E-08	1.0E-04	8.4E-09	4.0E-09	2.6E-09	1.7E-09	1.4E-09
		S	0.001	1.3E-08	1.0E-04	8.8E-09	4.2E-09	2.7E-09	1.8E-09	1.5E-09

a Dose coefficients for this element are based on age-specific biokinetic data

b The f_1 value for 1 to 15 year olds for Type F is 0.4

Table A.2.—(continued)

Nuclide	Physical half-life	Type	f_1 <1y	e(τ) 3 Months	f_1 ≥1y	e(τ) 1 Year	5 Years	10 Years	15 Years	Adult
Y-90m	3.19 h	M	0.001	7.2E-10	1.0E-04	5.7E-10	2.8E-10	1.8E-10	1.1E-10	9.5E-11
		S	0.001	7.5E-10	1.0E-04	6.0E-10	2.9E-10	1.9E-10	1.2E-10	1.0E-10
Y-91	58.5 d	M	0.001	3.9E-08	1.0E-04	3.0E-08	1.6E-08	1.1E-08	8.4E-09	7.1E-09
		S	0.001	4.3E-08	1.0E-04	3.4E-08	1.9E-08	1.3E-08	1.0E-08	8.9E-09
Y-91m	0.828 h	M	0.001	7.0E-11	1.0E-04	5.5E-11	2.9E-11	1.8E-11	1.2E-11	1.0E-11
		S	0.001	7.4E-11	1.0E-04	5.9E-11	3.1E-11	2.0E-11	1.4E-11	1.1E-11
Y-92	3.54 h	M	0.001	1.8E-09	1.0E-04	1.2E-09	5.3E-10	3.3E-10	2.0E-10	1.7E-10
		S	0.001	1.9E-09	1.0E-04	1.2E-09	5.5E-10	3.5E-10	2.1E-10	1.8E-10
Y-93	10.1 h	M	0.001	4.4E-09	1.0E-04	2.9E-09	1.3E-09	8.1E-10	4.7E-10	4.0E-10
		S	0.001	4.6E-09	1.0E-04	3.0E-09	1.4E-09	8.5E-10	5.0E-10	4.2E-10
Y-94	0.318 h	M	0.001	2.8E-10	1.0E-04	1.8E-10	8.1E-11	5.0E-11	3.1E-11	2.7E-11
		S	0.001	2.9E-10	1.0E-04	1.9E-10	8.4E-11	5.2E-11	3.3E-11	2.8E-11
Y-95	0.178 h	M	0.001	1.5E-10	1.0E-04	9.8E-11	4.4E-11	2.8E-11	1.8E-11	1.5E-11
		S	0.001	1.6E-10	1.0E-04	1.0E-10	4.5E-11	2.9E-11	1.8E-11	1.6E-11
Zirconium[a]										
Zr-86	16.5 h	F	0.020	2.4E-09	0.002	1.9E-09	9.5E-10	5.9E-10	3.4E-10	2.7E-10
		M	0.020	3.4E-09	0.002	2.6E-09	1.3E-09	8.4E-10	5.2E-10	4.2E-10
		S	0.020	3.5E-09	0.002	2.7E-09	1.4E-09	8.7E-10	5.4E-10	4.3E-10
Zr-88	83.4 d	F	0.020	6.9E-09	0.002	8.3E-09	5.6E-09	4.7E-09	3.6E-09	3.5E-09
		M	0.020	8.5E-09	0.002	7.8E-09	5.1E-09	3.6E-09	3.0E-09	2.6E-09
		S	0.020	1.3E-08	0.002	1.2E-08	7.7E-09	5.2E-09	4.3E-09	3.6E-09
Zr-89	3.27 d	F	0.020	2.6E-09	0.002	2.0E-09	9.9E-10	6.1E-10	3.6E-10	2.9E-10
		M	0.020	3.7E-09	0.002	2.8E-09	1.5E-09	9.6E-10	6.5E-10	5.2E-10
		S	0.020	3.9E-09	0.002	2.9E-09	1.5E-09	1.0E-09	6.8E-10	5.5E-10
Zr-93	1.53E+06 y	F	0.020	3.5E-09	0.002	4.8E-09	5.3E-09	9.7E-09	1.8E-08	2.5E-08
		M	0.020	3.3E-09	0.002	3.1E-09	2.8E-09	4.1E-09	7.5E-09	1.0E-08
		S	0.020	7.0E-09	0.002	6.4E-09	4.5E-09	3.3E-09	3.3E-09	3.3E-09
Zr-95	64.0 d	F	0.020	1.2E-08	0.002	1.1E-08	6.4E-09	4.2E-09	2.8E-09	2.5E-09
		M	0.020	2.0E-08	0.002	1.6E-08	9.7E-09	6.8E-09	5.9E-09	4.8E-09
		S	0.020	2.4E-08	0.002	1.9E-08	1.2E-08	8.3E-09	7.3E-09	5.9E-09
Zr-97	16.9 h	F	0.020	5.0E-09	0.002	3.4E-09	1.5E-09	9.1E-10	4.8E-10	3.9E-10
		M	0.020	7.8E-09	0.002	5.3E-09	2.8E-09	1.8E-09	1.1E-09	9.2E-10
		S	0.020	8.2E-09	0.002	5.6E-09	2.9E-09	1.9E-09	1.2E-09	8.9E-10
Niobium[a]										
Nb-88	0.238 h	F	0.020	1.8E-10	0.010	1.3E-10	6.3E-11	3.9E-11	2.4E-11	1.9E-11
		M	0.020	2.5E-10	0.010	1.8E-10	8.5E-11	5.3E-11	3.3E-11	2.7E-11
		S	0.020	2.6E-10	0.010	1.8E-10	8.7E-11	5.5E-11	3.5E-11	2.8E-11
Nb-89	2.03 h	F	0.020	7.0E-10	0.010	4.8E-10	2.2E-10	1.3E-10	7.4E-11	6.1E-11
		M	0.020	1.1E-09	0.010	7.6E-10	3.6E-10	2.2E-10	1.4E-10	1.1E-10
		S	0.020	1.2E-09	0.010	7.9E-10	3.7E-10	2.3E-10	1.5E-10	1.2E-10
Nb-89	1.10 h	F	0.020	4.0E-10	0.010	2.9E-10	1.4E-10	8.3E-11	4.8E-11	3.9E-11
		M	0.020	6.2E-10	0.010	4.3E-10	2.1E-10	1.3E-10	8.2E-11	6.8E-11
		S	0.020	6.4E-10	0.010	4.4E-10	2.1E-10	1.4E-10	8.6E-11	7.1E-11
Nb-90	14.6 h	F	0.020	3.5E-09	0.010	2.7E-09	1.3E-09	8.2E-10	4.7E-10	3.8E-10
		M	0.020	5.1E-09	0.010	3.9E-09	1.9E-09	1.3E-09	7.8E-10	6.3E-10
		S	0.020	5.3E-09	0.010	4.0E-09	2.0E-09	1.3E-09	8.1E-10	6.6E-10

a   Dose coefficients for this element are based on age-specific biokinetic data

Table A.2.—(*continued*)

Nuclide	Physical half-life	Type	f_1 <1y	$e(\tau)$ 3 Months	f_1 ≥1y	$e(\tau)$ 1 Year	5 Years	10 Years	15 Years	Adult
Nb-93m	13.6 y	F	0.020	1.8E-09	0.010	1.4E-09	7.0E-10	4.4E-10	2.7E-10	2.2E-10
		M	0.020	3.1E-09	0.010	2.4E-09	1.3E-09	8.2E-10	5.9E-10	5.1E-10
		S	0.020	7.4E-09	0.010	6.5E-09	4.0E-09	2.5E-09	1.9E-09	1.8E-09
Nb-94	2.03E+04 y	F	0.020	3.1E-08	0.010	2.7E-08	1.5E-08	1.0E-08	6.7E-09	5.8E-09
		M	0.020	4.3E-08	0.010	3.7E-08	2.3E-08	1.6E-08	1.3E-08	1.1E-08
		S	0.020	1.2E-07	0.010	1.2E-07	8.3E-08	5.8E-08	5.2E-08	4.9E-08
Nb-95	35.1 d	F	0.020	4.1E-09	0.010	3.1E-09	1.6E-09	1.2E-09	7.5E-10	5.7E-10
		M	0.020	6.8E-09	0.010	5.2E-09	3.1E-09	2.2E-09	1.9E-09	1.5E-09
		S	0.020	7.7E-09	0.010	5.9E-09	3.6E-09	2.5E-09	2.2E-09	1.8E-09
Nb-95m	3.61 d	F	0.020	2.3E-09	0.010	1.6E-09	7.0E-10	4.2E-10	2.4E-10	2.0E-10
		M	0.020	4.3E-09	0.010	3.1E-09	1.7E-09	1.2E-09	1.0E-09	7.9E-10
		S	0.020	4.6E-09	0.010	3.4E-09	1.9E-09	1.3E-09	1.1E-09	8.8E-10
Nb-96	23.3 h	F	0.020	3.1E-09	0.010	2.4E-09	1.2E-09	7.3E-10	4.2E-10	3.4E-10
		M	0.020	4.7E-09	0.010	3.6E-09	1.8E-09	1.2E-09	7.8E-10	5.3E-10
		S	0.020	4.9E-09	0.010	3.7E-09	1.9E-09	1.2E-09	8.3E-10	6.6E-10
Nb-97	1.20 h	F	0.020	2.2E-10	0.010	1.5E-10	6.8E-11	4.2E-11	2.5E-11	2.1E-11
		M	0.020	3.7E-10	0.010	2.5E-10	1.2E-10	7.7E-11	5.2E-11	4.3E-11
		S	0.020	3.8E-10	0.010	2.6E-10	1.2E-10	8.1E-11	5.5E-11	4.5E-11
Nb-98	0.858 h	F	0.020	3.4E-10	0.010	2.4E-10	1.1E-10	6.9E-11	4.1E-11	3.3E-11
		M	0.020	5.2E-10	0.010	3.6E-10	1.7E-10	1.1E-10	6.8E-11	5.6E-11
		S	0.020	5.3E-10	0.010	3.7E-10	1.8E-10	1.1E-10	7.1E-11	5.8E-11
Molybdenum*										
Mo-90	5.67 h	F	1.000	1.2E-09	0.800	1.1E-09	5.3E-10	3.2E-10	1.9E-10	1.5E-10
		M	0.200	2.6E-09	0.100	2.0E-09	9.9E-10	6.5E-10	4.2E-10	3.4E-10
		S	0.020	2.8E-09	0.010	2.1E-09	1.1E-09	6.9E-10	4.5E-10	3.6E-10
Mo-93	3.50E+03 y	F	1.000	3.1E-09	0.800	2.6E-09	1.7E-09	1.3E-09	1.1E-09	1.0E-09
		M	0.200	2.2E-09	0.100	1.8E-09	1.1E-09	7.9E-10	6.6E-10	5.9E-10
		S	0.020	6.0E-09	0.010	5.8E-09	4.0E-09	2.8E-09	2.4E-09	2.3E-09
Mo-93m	6.85 h	F	1.000	7.3E-10	0.800	6.4E-10	3.3E-10	2.0E-10	1.2E-10	9.6E-11
		M	0.200	1.2E-09	0.100	9.7E-10	5.0E-10	3.2E-10	2.0E-10	1.6E-10
		S	0.020	1.3E-09	0.010	1.0E-09	5.2E-10	3.4E-10	2.1E-10	1.7E-10
Mo-99	2.75 d	F	1.000	2.3E-09	0.800	1.7E-09	7.7E-10	4.7E-10	2.6E-10	2.2E-10
		M	0.200	6.0E-09	0.100	4.4E-09	2.2E-09	1.5E-09	1.1E-09	8.9E-10
		S	0.020	6.9E-09	0.010	4.8E-09	2.4E-09	1.7E-09	1.2E-09	9.9E-10
Mo-101	0.244 h	F	1.000	1.4E-10	0.800	9.7E-11	4.4E-11	2.8E-11	1.7E-11	1.4E-11
		M	0.200	2.2E-10	0.100	1.5E-10	7.0E-11	4.5E-11	3.0E-11	2.5E-11
		S	0.020	2.3E-10	0.010	1.6E-10	7.2E-11	4.7E-11	3.1E-11	2.6E-11
Technetium*										
Tc-93	2.75 h	F	1.000	2.4E-10	0.800	2.1E-10	1.1E-10	6.7E-11	4.0E-11	3.2E-11
		M	0.200	2.7E-10	0.100	2.3E-10	1.2E-10	7.5E-11	4.4E-11	3.5E-11
		S	0.020	2.8E-10	0.010	2.3E-10	1.2E-10	7.6E-11	4.5E-11	3.5E-11
Tc-93m	0.725 h	F	1.000	1.2E-10	0.800	9.8E-11	4.9E-11	2.9E-11	1.8E-11	1.4E-11
		M	0.200	1.4E-10	0.100	1.1E-10	5.4E-11	3.4E-11	2.1E-11	1.7E-11
		S	0.020	1.4E-10	0.010	1.1E-10	5.4E-11	3.4E-11	2.1E-11	1.7E-11
Tc-94	4.88 h	F	1.000	8.9E-10	0.800	7.5E-10	3.9E-10	2.3E-10	1.4E-10	1.1E-10
		M	0.200	9.8E-10	0.100	8.1E-10	4.2E-10	2.6E-10	1.6E-10	1.2E-10
		S	0.020	9.9E-10	0.010	8.2E-10	4.3E-10	2.7E-10	1.6E-10	1.3E-10
Tc-94m	0.867 h	F	1.000	4.8E-10	0.800	3.4E-10	1.6E-10	8.6E-11	5.2E-11	4.1E-11
		M	0.200	4.4E-10	0.100	3.0E-10	1.4E-10	8.8E-11	5.5E-11	4.5E-11
		S	0.020	4.3E-10	0.010	3.0E-10	1.4E-10	8.8E-11	5.6E-11	4.6E-11

a Dose coefficients for this element are based on age-specific biokinetic data

Table A.2.—*(continued)*

Nuclide	Physical half-life	Type	f_1 <1y	$e(\tau)$ 3 Months	f_1 ≥1y	$e(\tau)$ 1 Year	5 Years	10 Years	15 Years	Adult
Tc-95	20.0 h	F	1.000	7.5E-10	0.800	6.3E-10	3.3E-10	2.0E-10	1.2E-10	9.6E-11
		M	0.200	8.3E-10	0.100	6.9E-10	3.6E-10	2.2E-10	1.3E-10	1.0E-10
		S	0.020	8.5E-10	0.010	7.0E-10	3.6E-10	2.3E-10	1.4E-10	1.1E-10
Tc-95m	61.0 d	F	1.000	2.4E-09	0.800	1.8E-09	9.3E-10	5.7E-10	3.6E-10	2.9E-10
		M	0.200	4.9E-09	0.100	4.0E-09	2.3E-09	1.5E-09	1.1E-09	8.8E-10
		S	0.020	6.0E-09	0.010	5.0E-09	2.7E-09	1.8E-09	1.5E-09	1.2E-09
Tc-96	4.28 d	F	1.000	4.2E-09	0.800	3.4E-09	1.8E-09	1.1E-09	7.0E-10	5.7E-10
		M	0.200	4.7E-09	0.100	3.9E-09	2.1E-09	1.3E-09	8.6E-10	6.8E-10
		S	0.020	4.8E-09	0.010	3.9E-09	2.1E-09	1.4E-09	8.9E-10	7.0E-10
Tc-96m	0.858 h	F	1.000	5.3E-11	0.800	4.1E-11	2.1E-11	1.3E-11	7.7E-12	6.2E-12
		M	0.200	5.6E-11	0.100	4.4E-11	2.3E-11	1.4E-11	9.3E-12	7.4E-12
		S	0.020	5.7E-11	0.010	4.4E-11	2.3E-11	1.5E-11	9.5E-12	7.5E-12
Tc-97	2.60E+06 y	F	1.000	5.2E-10	0.800	3.7E-10	1.7E-10	9.4E-11	5.6E-11	4.3E-11
		M	0.200	1.2E-09	0.100	1.0E-09	5.7E-10	3.6E-10	2.8E-10	2.2E-10
		S	0.020	5.0E-09	0.010	4.8E-09	3.3E-09	2.2E-09	1.9E-09	1.8E-09
Tc-97m	87.0 d	F	1.000	3.4E-09	0.800	2.3E-09	9.8E-10	5.6E-10	3.0E-10	2.7E-10
		M	0.200	1.3E-08	0.100	1.0E-08	6.1E-09	4.4E-09	4.1E-09	3.2E-09
		S	0.020	1.6E-08	0.010	1.3E-08	7.8E-09	5.7E-09	5.2E-09	4.1E-09
Tc-98	4.20E+06 y	F	1.000	1.0E-08	0.800	6.8E-09	3.2E-09	1.9E-09	1.2E-09	9.7E-10
		M	0.200	3.5E-08	0.100	2.9E-08	1.7E-08	1.2E-08	1.0E-08	8.3E-09
		S	0.020	1.1E-07	0.010	1.1E-07	7.6E-08	5.4E-08	4.8E-08	4.5E-08
Tc-99	2.13E+05 y	F	1.000	4.0E-09	0.800	2.5E-09	1.0E-09	5.9E-10	3.6E-10	2.9E-10
		M	0.200	1.7E-08	0.100	1.3E-08	8.0E-09	5.7E-09	5.0E-09	4.0E-09
		S	0.020	4.1E-08	0.010	3.7E-08	2.4E-08	1.7E-08	1.5E-08	1.3E-08
Tc-99m	6.02 h	F	1.000	1.2E-10	0.800	8.7E-11	4.1E-11	2.4E-11	1.5E-11	1.2E-11
		M	0.200	1.3E-10	0.100	9.9E-11	5.1E-11	3.4E-11	2.4E-11	1.9E-11
		S	0.020	1.3E-10	0.010	1.0E-10	5.2E-11	3.5E-11	2.5E-11	2.0E-11
Tc-101	0.237 h	F	1.000	8.5E-11	0.800	5.6E-11	2.5E-11	1.6E-11	9.7E-12	8.2E-12
		M	0.200	1.1E-10	0.100	7.1E-11	3.2E-11	2.1E-11	1.4E-11	1.2E-11
		S	0.020	1.1E-10	0.010	7.3E-11	3.3E-11	2.2E-11	1.4E-11	1.2E-11
Tc-104	0.303 h	F	1.000	2.7E-10	0.800	1.8E-10	8.0E-11	4.6E-11	2.8E-11	2.3E-11
		M	0.200	2.9E-10	0.100	1.9E-10	8.6E-11	5.4E-11	3.3E-11	2.8E-11
		S	0.020	2.9E-10	0.010	1.9E-10	8.7E-11	5.4E-11	3.4E-11	2.9E-11
Ruthenium[a]										
Ru-94	0.863 h	F	0.100	2.5E-10	0.050	1.9E-10	9.0E-11	5.4E-11	3.1E-11	2.5E-11
		M	0.100	3.8E-10	0.050	2.8E-10	1.3E-10	8.4E-11	5.2E-11	4.2E-11
		S	0.020	4.0E-10	0.010	2.9E-10	1.4E-10	8.7E-11	5.4E-11	4.4E-11
Ru-97	2.90 d	F	0.100	5.5E-10	0.050	4.4E-10	2.2E-10	1.3E-10	7.7E-11	6.2E-11
		M	0.100	7.7E-10	0.050	6.1E-10	3.1E-10	2.0E-10	1.3E-10	1.0E-10
		S	0.020	8.1E-10	0.010	6.3E-10	3.3E-10	2.1E-10	1.4E-10	1.1E-10
Ru-103	39.3 d	F	0.100	4.2E-09	0.050	3.0E-09	1.5E-09	9.3E-10	5.6E-10	4.8E-10
		M	0.100	1.1E-08	0.050	8.4E-09	5.0E-09	3.5E-09	3.0E-09	2.4E-09
		S	0.020	1.3E-08	0.010	1.0E-08	6.0E-09	4.2E-09	3.7E-09	3.0E-09
Ru-105	4.44 h	F	0.100	7.1E-10	0.050	5.1E-10	2.3E-10	1.4E-10	7.9E-11	6.5E-11
		M	0.100	1.3E-09	0.050	9.2E-10	4.5E-10	3.0E-10	2.0E-10	1.7E-10
		S	0.020	1.4E-09	0.010	9.8E-10	4.8E-10	3.2E-10	2.2E-10	1.8E-10
Ru-106	1.01 y	F	0.100	7.2E-08	0.050	5.4E-08	2.6E-08	1.6E-08	9.2E-09	7.9E-09
		M	0.100	1.4E-07	0.050	1.1E-07	6.4E-08	4.1E-08	3.1E-08	2.8E-08
		S	0.020	2.6E-07	0.010	2.3E-07	1.4E-07	9.1E-08	7.1E-08	6.6E-08

a   Dose coefficients for this element are based on age-specific biokinetic data

Table A.2.—(continued)

Nuclide	Physical half-life	Type	f₁ <1y	e(τ) 3 Months	f₁ ≥1y	e(τ) 1 Year	5 Years	10 Years	15 Years	Adult

Nuclide	Physical half-life	Type	f_1 <1y	$e(\tau)$ 3 Months	f_1 ≥1y	$e(\tau)$ 1 Year	5 Years	10 Years	15 Years	Adult
Rhodium										
Rh-99	16.0 d	F	0.100	2.6E-09	0.050	2.0E-09	9.9E-10	6.2E-10	3.8E-10	3.2E-10
		M	0.100	4.5E-09	0.050	3.5E-09	2.0E-09	1.3E-09	9.6E-10	7.7E-10
		S	0.100	4.9E-09	0.050	3.8E-09	2.2E-09	1.3E-09	1.1E-09	8.7E-10
Rh-99m	4.70 h	F	0.100	2.4E-10	0.050	2.0E-10	1.0E-10	6.1E-11	3.5E-11	2.8E-11
		M	0.100	3.1E-10	0.050	2.5E-10	1.3E-10	8.0E-11	4.9E-11	3.9E-11
		S	0.100	3.2E-10	0.050	2.6E-10	1.3E-10	8.2E-11	5.1E-11	4.0E-11
Rh-100	20.8 h	F	0.100	2.1E-09	0.050	1.8E-09	9.1E-10	5.6E-10	3.3E-10	2.6E-10
		M	0.100	2.7E-09	0.050	2.2E-09	1.1E-09	7.1E-10	4.3E-10	3.4E-10
		S	0.100	2.8E-09	0.050	2.2E-09	1.2E-09	7.3E-10	4.4E-10	3.5E-10
Rh-101	3.20 y	F	0.100	7.4E-09	0.050	6.1E-09	3.5E-09	2.3E-09	1.5E-09	1.4E-09
		M	0.100	9.8E-09	0.050	8.0E-09	4.9E-09	3.4E-09	2.8E-09	2.3E-09
		S	0.100	1.9E-08	0.050	1.7E-08	1.1E-08	7.4E-09	6.2E-09	5.4E-09
Rh-101m	4.34 d	F	0.100	8.4E-10	0.050	6.6E-10	3.3E-10	2.0E-10	1.2E-10	9.7E-11
		M	0.100	1.3E-09	0.050	9.8E-10	5.2E-10	3.5E-10	2.5E-10	1.9E-10
		S	0.100	1.3E-09	0.050	1.0E-09	5.5E-10	3.7E-10	2.7E-10	2.1E-10
Rh-102	2.90 y	F	0.100	3.3E-08	0.050	2.8E-08	1.7E-08	1.1E-08	7.9E-09	7.3E-09
		M	0.100	3.0E-08	0.050	2.5E-08	1.5E-08	1.0E-08	7.9E-09	6.9E-09
		S	0.100	5.4E-08	0.050	5.0E-08	3.5E-08	2.4E-08	2.0E-08	1.7E-08
Rh-102m	207 d	F	0.100	1.2E-08	0.050	8.7E-09	4.4E-09	2.7E-09	1.7E-09	1.5E-09
		M	0.100	2.0E-08	0.050	1.6E-08	9.0E-09	6.0E-09	4.7E-09	4.0E-09
		S	0.100	3.0E-08	0.050	2.5E-08	1.5E-08	1.0E-08	8.2E-09	7.1E-09
Rh-103m	0.935 h	F	0.100	8.6E-12	0.050	5.9E-12	2.7E-12	1.6E-12	1.0E-12	8.6E-13
		M	0.100	1.9E-11	0.050	1.2E-11	6.3E-12	4.0E-12	3.0E-12	2.5E-12
		S	0.100	2.0E-11	0.050	1.3E-11	6.7E-12	4.3E-12	3.2E-12	2.7E-12
Rh-105	1.47 d	F	0.100	1.0E-09	0.050	6.9E-10	3.0E-10	1.8E-10	9.6E-11	8.2E-11
		M	0.100	2.2E-09	0.050	1.6E-09	7.4E-10	5.2E-10	4.1E-10	3.2E-10
		S	0.100	2.4E-09	0.050	1.7E-09	8.0E-10	5.6E-10	4.5E-10	3.5E-10
Rh-106m	2.20 h	F	0.100	5.7E-10	0.050	4.5E-10	2.2E-10	1.4E-10	8.0E-11	6.5E-11
		M	0.100	8.2E-10	0.050	6.3E-10	3.2E-10	2.0E-10	1.3E-10	1.1E-10
		S	0.100	8.5E-10	0.050	6.5E-10	3.3E-10	2.1E-10	1.4E-10	1.1E-10
Rh-107	0.362 h	F	0.100	8.9E-11	0.050	5.9E-11	2.6E-11	1.7E-11	1.0E-11	9.0E-12
		M	0.100	1.4E-10	0.050	9.3E-11	4.2E-11	2.8E-11	1.9E-11	1.6E-11
		S	0.100	1.5E-10	0.050	9.7E-11	4.4E-11	2.9E-11	1.9E-11	1.7E-11
Palladium										
Pd-100	3.63 d	F	0.050	3.9E-09	0.005	3.0E-09	1.5E-09	9.7E-10	5.8E-10	4.7E-10
		M	0.050	5.2E-09	0.005	4.0E-09	2.2E-09	1.4E-09	9.9E-10	8.0E-10
		S	0.050	5.3E-09	0.005	4.1E-09	2.2E-09	1.5E-09	1.0E-09	8.5E-10
Pd-101	8.27 h	F	0.050	3.6E-10	0.005	2.9E-10	1.4E-10	8.6E-11	4.9E-11	3.9E-11
		M	0.050	4.8E-10	0.005	3.8E-10	1.9E-10	1.2E-10	7.5E-11	5.9E-11
		S	0.050	5.0E-10	0.005	3.9E-10	2.0E-10	1.2E-10	7.8E-11	6.2E-11
Pd-103	17.0 d	F	0.050	9.7E-10	0.005	6.5E-10	3.0E-10	1.9E-10	1.1E-10	8.9E-11
		M	0.050	2.3E-09	0.005	1.6E-09	9.0E-10	5.9E-10	4.5E-10	3.8E-10
		S	0.050	2.5E-09	0.005	1.8E-09	1.0E-09	6.8E-10	5.3E-10	4.5E-10
Pd-107	6.50E+06 y	F	0.050	2.6E-10	0.005	1.8E-10	8.2E-11	5.2E-11	3.1E-11	2.5E-11
		M	0.050	6.5E-10	0.005	5.0E-10	2.6E-10	1.5E-10	1.0E-10	8.5E-11
		S	0.050	2.2E-09	0.005	2.0E-09	1.3E-09	7.8E-10	6.2E-10	5.9E-10
Pd-109	13.4 h	F	0.050	1.5E-09	0.005	9.9E-10	4.2E-10	2.6E-10	1.4E-10	1.2E-10
		M	0.050	2.6E-09	0.005	1.8E-09	8.8E-10	5.9E-10	4.3E-10	3.4E-10
		S	0.050	2.7E-09	0.005	1.9E-09	9.3E-10	6.3E-10	4.6E-10	3.7E-10

Table A.2.—(continued)

Nuclide	Physical half-life	Type	f_1 <1y	$e(\tau)$ 3 Months	f_1 ≥1y	$e(\tau)$ 1 Year	5 Years	10 Years	15 Years	Adult
Silver[a]										
Ag-102	0.215 h	F	0.100	1.2E-10	0.050	8.6E-11	4.2E-11	2.6E-11	1.5E-11	1.3E-11
		M	0.100	1.6E-10	0.050	1.1E-10	5.5E-11	3.4E-11	2.1E-11	1.7E-11
		S	0.020	1.6E-10	0.010	1.2E-10	5.6E-11	3.5E-11	2.2E-11	1.8E-11
Ag-103	1.09 h	F	0.100	1.4E-10	0.050	1.0E-10	4.9E-11	3.0E-11	1.8E-11	1.4E-11
		M	0.100	2.2E-10	0.050	1.6E-10	7.6E-11	4.8E-11	3.2E-11	2.6E-11
		S	0.020	2.3E-10	0.010	1.6E-10	7.9E-11	5.1E-11	3.3E-11	2.7E-11
Ag-104	1.15 h	F	0.100	2.3E-10	0.050	1.9E-10	9.8E-11	5.9E-11	3.5E-11	2.8E-11
		M	0.100	2.9E-10	0.050	2.3E-10	1.2E-10	7.4E-11	4.5E-11	3.6E-11
		S	0.020	2.9E-10	0.010	2.4E-10	1.2E-10	7.6E-11	4.6E-11	3.7E-11
Ag-104m	0.558 h	F	0.100	1.6E-10	0.050	1.1E-10	5.5E-11	3.4E-11	2.0E-11	1.6E-11
		M	0.100	2.3E-10	0.050	1.6E-10	7.7E-11	4.8E-11	3.0E-11	2.5E-11
		S	0.020	2.4E-10	0.010	1.7E-10	8.0E-11	5.0E-11	3.1E-11	2.6E-11
Ag-105	41.0 d	F	0.100	3.9E-09	0.050	3.4E-09	1.7E-09	1.0E-09	6.4E-10	5.4E-10
		M	0.100	4.5E-09	0.050	3.5E-09	2.0E-09	1.3E-09	9.0E-10	7.3E-10
		S	0.020	4.5E-09	0.010	3.6E-09	2.1E-09	1.3E-09	1.0E-09	8.1E-10
Ag-106	0.399 h	F	0.100	9.4E-11	0.050	6.4E-11	2.9E-11	1.8E-11	1.1E-11	9.1E-12
		M	0.100	1.4E-10	0.050	9.5E-11	4.4E-11	2.8E-11	1.8E-11	1.5E-11
		S	0.020	1.5E-10	0.010	9.9E-11	4.5E-11	2.9E-11	1.9E-11	1.6E-11
Ag-106m	8.41 d	F	0.100	7.7E-09	0.050	6.1E-09	3.2E-09	2.1E-09	1.3E-09	1.1E-09
		M	0.100	7.2E-09	0.050	5.8E-09	3.2E-09	2.1E-09	1.4E-09	1.1E-09
		S	0.020	7.0E-09	0.010	5.7E-09	3.2E-09	2.1E-09	1.4E-09	1.1E-09
Ag-108m	1.27E+02 y	F	0.100	3.5E-08	0.050	2.8E-08	1.6E-08	1.0E-08	6.9E-09	6.1E-09
		M	0.100	3.3E-08	0.050	2.7E-08	1.7E-08	1.1E-08	8.6E-09	7.4E-09
		S	0.020	8.9E-08	0.010	8.7E-08	6.2E-08	4.4E-08	3.9E-08	3.7E-08
Ag-110m	250 d	F	0.100	3.5E-08	0.050	2.8E-08	1.5E-08	9.7E-09	6.3E-09	5.5E-09
		M	0.100	3.5E-08	0.050	2.8E-08	1.7E-08	1.2E-08	9.2E-09	7.6E-09
		S	0.020	4.6E-08	0.010	4.1E-08	2.6E-08	1.8E-08	1.5E-06	1.2E-08
Ag-111	7.45 d	F	0.100	4.8E-09	0.050	3.2E-09	1.4E-09	8.8E-10	4.8E-10	4.0E-10
		M	0.100	9.2E-09	0.050	6.6E-09	3.5E-09	2.4E-09	1.9E-09	1.5E-09
		S	0.020	9.9E-09	0.010	7.1E-09	3.8E-09	2.7E-09	2.1E-09	1.7E-09
Ag-112	3.12 h	F	0.100	9.8E-10	0.050	6.4E-10	2.8E-10	1.7E-10	9.1E-11	7.6E-11
		M	0.100	1.7E-09	0.050	1.1E-09	5.1E-10	3.2E-10	2.0E-10	1.6E-10
		S	0.020	1.8E-09	0.010	1.2E-09	5.4E-10	3.4E-10	2.1E-10	1.7E-10
Ag-115	0.333 h	F	0.100	1.6E-10	0.050	1.0E-10	4.6E-11	2.9E-11	1.7E-11	1.5E-11
		M	0.100	2.5E-10	0.050	1.7E-10	7.6E-11	4.9E-11	3.2E-11	2.7E-11
		S	0.020	2.7E-10	0.010	1.7E-10	8.0E-11	5.2E-11	3.4E-11	2.9E-11
Cadmium										
Cd-104	0.961 h	F	0.100	2.0E-10	0.050	1.7E-10	8.7E-11	5.2E-11	3.1E-11	2.4E-11
		M	0.100	2.6E-10	0.050	2.1E-10	1.1E-10	6.9E-11	4.2E-11	3.4E-11
		S	0.100	2.7E-10	0.050	2.2E-10	1.1E-10	7.0E-11	4.4E-11	3.5E-11
Cd-107	6.49 h	F	0.100	2.3E-10	0.050	1.7E-10	7.4E-11	4.6E-11	2.5E-11	2.1E-11
		M	0.100	5.2E-10	0.050	3.7E-10	2.0E-10	1.3E-10	8.8E-11	8.3E-11
		S	0.100	5.5E-10	0.050	3.9E-10	2.1E-10	1.4E-10	9.7E-11	7.7E-11
Cd-109	1.27 y	F	0.100	4.5E-08	0.050	3.7E-08	2.1E-08	1.4E-08	9.3E-09	8.1E-09
		M	0.100	3.0E-08	0.050	2.3E-08	1.4E-08	9.5E-09	7.8E-09	6.6E-09
		S	0.100	2.7E-08	0.050	2.1E-08	1.3E-08	8.9E-09	7.6E-09	6.2E-09
Cd-113	9.30E+15 y	F	0.100	2.6E-07	0.050	2.4E-07	1.7E-07	1.4E-07	1.2E-07	1.2E-07
		M	0.100	1.2E-07	0.050	1.0E-07	7.6E-08	6.1E-08	5.7E-08	5.5E-08
		S	0.100	7.8E-08	0.050	5.8E-08	4.1E-08	3.0E-08	2.7E-08	2.6E-08

a   Dose coefficients for this element are based on age-specific biokinetic data

Table A.2.—(continued)

Nuclide	Physical half-life	Type	f_1 <1y	$e(\tau)$ 3 Months	f_1 ≥1y	$e(\tau)$ 1 Year	5 Years	10 Years	15 Years	Adult
Cd-113m	13.6 y	F	0.100	3.0E-07	0.050	2.7E-07	1.8E-07	1.3E-07	1.1E-07	1.1E-07
		M	0.100	1.4E-07	0.050	1.2E-07	8.1E-08	6.0E-08	5.3E-08	5.2E-08
		S	0.100	1.1E-07	0.050	8.4E-08	5.5E-08	3.9E-08	3.3E-08	3.1E-08
Cd-115	2.23 d	F	0.100	4.0E-09	0.050	2.6E-09	1.2E-09	7.5E-10	4.3E-10	3.5E-10
		M	0.100	6.7E-09	0.050	4.8E-09	2.4E-09	1.7E-09	1.2E-09	9.8E-10
		S	0.100	7.2E-09	0.050	5.1E-09	2.6E-09	1.8E-09	1.3E-09	1.1E-09
Cd-115m	44.6 d	F	0.100	4.6E-08	0.050	3.2E-08	1.5E-08	1.0E-08	6.4E-09	5.3E-09
		M	0.100	4.0E-08	0.050	2.5E-08	1.4E-08	9.4E-09	7.3E-09	6.2E-09
		S	0.100	3.9E-08	0.050	3.0E-08	1.7E-08	1.1E-08	8.9E-09	7.7E-09
Cd-117	2.49 h	F	0.100	7.4E-10	0.050	5.2E-10	2.4E-10	1.5E-10	8.1E-11	6.7E-11
		M	0.100	1.3E-09	0.050	9.3E-10	4.5E-10	2.9E-10	2.0E-10	1.6E-10
		S	0.100	1.4E-09	0.050	9.8E-10	4.8E-10	3.1E-10	2.1E-10	1.7E-10
Cd-117m	3.36 h	F	0.100	8.9E-10	0.050	6.7E-10	3.3E-10	2.0E-10	1.1E-10	9.4E-11
		M	0.100	1.5E-09	0.050	1.1E-09	5.5E-10	3.6E-10	2.4E-10	2.0E-10
		S	0.100	1.5E-09	0.050	1.1E-09	5.7E-10	3.8E-10	2.6E-10	2.1E-10

Indium

Nuclide	Physical half-life	Type	f_1 <1y	$e(\tau)$ 3 Months	f_1 ≥1y	$e(\tau)$ 1 Year	5 Years	10 Years	15 Years	Adult
In-109	4.20 h	F	0.040	2.6E-10	0.020	2.1E-10	1.0E-10	6.3E-11	3.6E-11	2.9E-11
		M	0.040	3.3E-10	0.020	2.6E-10	1.3E-10	8.4E-11	5.3E-11	4.2E-11
In-110	4.90 h	F	0.040	8.2E-10	0.020	7.1E-10	3.7E-10	2.3E-10	1.3E-10	1.1E-10
		M	0.040	9.9E-10	0.020	8.3E-10	4.4E-10	2.7E-10	1.6E-10	1.3E-10
In-110	1.15 h	F	0.040	3.0E-10	0.020	2.1E-10	9.9E-11	6.0E-11	3.5E-11	2.8E-11
		M	0.040	4.5E-10	0.020	3.1E-10	1.5E-10	9.2E-11	5.8E-11	4.7E-11
In-111	2.83 d	F	0.040	1.2E-09	0.020	8.6E-10	4.2E-10	2.6E-10	1.5E-10	1.3E-10
		M	0.040	1.5E-09	0.020	1.2E-09	6.2E-10	4.1E-10	2.9E-10	2.3E-10
In-112	0.240 h	F	0.040	4.4E-11	0.020	3.0E-11	1.3E-11	8.7E-12	5.4E-12	4.7E-12
		M	0.040	6.5E-11	0.020	4.4E-11	2.0E-11	1.3E-11	8.7E-12	7.4E-12
In-113m	1.66 h	F	0.040	1.0E-10	0.020	7.0E-11	3.2E-11	2.0E-11	1.2E-11	9.7E-12
		M	0.040	1.6E-10	0.020	1.1E-10	5.5E-11	3.6E-11	2.4E-11	2.0E-11
In-114m	49.5 d	F	0.040	1.2E-07	0.020	7.7E-08	3.4E-08	1.9E-08	1.1E-08	9.3E-09
		M	0.040	4.8E-08	0.020	3.3E-08	1.6E-08	1.0E-08	7.8E-09	6.1E-09
In-115	5.10E+15 y	F	0.040	8.3E-07	0.020	7.8E-07	5.5E-07	5.0E-07	4.2E-07	3.9E-07
		M	0.040	3.0E-07	0.020	2.8E-07	2.1E-07	1.9E-07	1.7E-07	1.6E-07
In-115m	4.49 h	F	0.040	2.8E-10	0.020	1.9E-10	8.4E-11	5.1E-11	2.8E-11	2.4E-11
		M	0.040	4.7E-10	0.020	3.3E-10	1.6E-10	1.0E-10	7.2E-11	5.9E-11
In-116m	0.902 h	F	0.040	2.5E-10	0.020	1.9E-10	9.2E-11	5.7E-11	3.4E-11	2.8E-11
		M	0.040	3.6E-10	0.020	2.7E-10	1.3E-10	8.5E-11	5.6E-11	4.5E-11
In-117	0.730 h	F	0.040	1.4E-10	0.020	9.7E-11	4.5E-11	2.8E-11	1.7E-11	1.5E-11
		M	0.040	2.3E-10	0.020	1.6E-10	7.5E-11	5.0E-11	3.5E-11	2.9E-11
In-117m	1.94 h	F	0.040	3.4E-10	0.020	2.3E-10	1.0E-10	6.2E-11	3.5E-11	2.9E-11
		M	0.040	6.0E-10	0.020	4.0E-10	1.9E-10	1.3E-10	8.7E-11	7.2E-11
In-119m	0.300 h	F	0.040	1.2E-10	0.020	7.3E-11	3.1E-11	2.0E-11	1.2E-11	1.0E-11
		M	0.040	1.8E-10	0.020	1.1E-10	4.9E-11	3.2E-11	2.0E-11	1.7E-11

Tin

Nuclide	Physical half-life	Type	f_1 <1y	$e(\tau)$ 3 Months	f_1 ≥1y	$e(\tau)$ 1 Year	5 Years	10 Years	15 Years	Adult
Sn-110	4.00 h	F	0.040	1.0E-09	0.020	7.6E-10	3.6E-10	2.2E-10	1.2E-10	9.9E-11
		M	0.040	1.5E-09	0.020	1.1E-09	5.1E-10	3.2E-10	1.9E-10	1.6E-10

Table A.2.—(continued)

Nuclide	Physical half-life	Type	f_1 <1y	$e(\tau)$ 3 Months	f_1 ≥1y	$e(\tau)$ 1 Year	5 Years	10 Years	15 Years	Adult
Sn-111	0.588 h	F	0.040	7.7E-11	0.020	5.4E-11	2.6E-11	1.6E-11	9.4E-12	7.8E-12
		M	0.040	1.1E-10	0.020	8.0E-11	3.8E-11	2.5E-11	1.6E-11	1.3E-11
Sn-113	115 d	F	0.040	5.1E-09	0.020	3.7E-09	1.8E-09	1.1E-09	6.4E-10	5.4E-10
		M	0.040	1.3E-08	0.020	1.0E-08	5.8E-09	4.0E-09	3.2E-09	2.7E-09
Sn-117m	13.6 d	F	0.040	3.3E-09	0.020	2.2E-09	1.0E-09	6.1E-10	3.4E-10	2.8E-10
		M	0.040	1.0E-08	0.020	7.7E-09	4.6E-09	3.4E-09	3.1E-09	2.4E-09
Sn-119m	293 d	F	0.040	3.0E-09	0.020	2.2E-09	1.0E-09	6.0E-10	3.4E-10	2.8E-10
		M	0.040	1.0E-08	0.020	7.9E-09	4.7E-09	3.1E-09	2.6E-09	2.2E-09
Sn-121	1.13 d	F	0.040	7.7E-10	0.020	5.0E-10	2.2E-10	1.3E-10	7.0E-11	6.0E-11
		M	0.040	1.5E-09	0.020	1.1E-09	5.1E-10	3.6E-10	2.9E-10	2.3E-10
Sn-121m	55.0 y	F	0.040	6.9E-09	0.020	5.4E-09	2.8E-09	1.6E-09	9.4E-10	8.0E-10
		M	0.040	1.9E-08	0.020	1.5E-08	9.2E-09	6.4E-09	5.5E-09	4.5E-09
Sn-123	129 d	F	0.040	1.4E-08	0.020	9.9E-09	4.5E-09	2.6E-09	1.4E-09	1.2E-09
		M	0.040	4.0E-08	0.020	3.1E-08	1.8E-08	1.2E-08	9.5E-09	8.1E-09
Sn-123m	0.668 h	F	0.040	1.4E-10	0.020	8.9E-11	3.9E-11	2.5E-11	1.5E-11	1.3E-11
		M	0.040	2.3E-10	0.020	1.5E-10	7.0E-11	4.6E-11	3.2E-11	2.7E-11
Sn-125	9.64 d	F	0.040	1.2E-08	0.020	8.0E-09	3.5E-09	2.0E-09	1.1E-09	8.9E-10
		M	0.040	2.1E-08	0.020	1.5E-08	7.6E-09	5.0E-09	3.6E-09	3.1E-09
Sn-126	1.00E+05 y	F	0.040	7.3E-08	0.020	5.9E-08	3.2E-08	2.0E-08	1.3E-08	1.1E-08
		M	0.040	1.2E-07	0.020	1.0E-07	6.2E-08	4.1E-08	3.3E-08	2.8E-08
Sn-127	2.10 h	F	0.040	6.6E-10	0.020	4.7E-10	2.3E-10	1.4E-10	7.9E-11	6.5E-11
		M	0.040	1.0E-09	0.020	7.4E-10	3.7E-10	2.4E-10	1.6E-10	1.3E-10
Sn-128	0.985 h	F	0.040	5.1E-10	0.020	3.6E-10	1.7E-10	1.0E-10	6.1E-11	5.0E-11
		M	0.040	8.0E-10	0.020	5.5E-10	2.7E-10	1.7E-10	1.1E-10	9.2E-11
Antimony[a]										
Sb-115	0.530 h	F	0.200	8.1E-11	0.100	5.9E-11	2.8E-11	1.7E-11	1.0E-11	8.5E-12
		M	0.020	1.2E-10	0.010	8.3E-11	4.0E-11	2.5E-11	1.6E-11	1.3E-11
		S	0.020	1.2E-10	0.010	8.6E-11	4.1E-11	2.6E-11	1.7E-11	1.4E-11
Sb-116	0.263 h	F	0.200	8.4E-11	0.100	6.2E-11	3.0E-11	1.9E-11	1.1E-11	9.1E-12
		M	0.020	1.1E-10	0.010	8.2E-11	4.0E-11	2.5E-11	1.5E-11	1.3E-11
		S	0.020	1.2E-10	0.010	8.5E-11	4.1E-11	2.6E-11	1.6E-11	1.3E-11
Sb-116m	1.00 h	F	0.200	2.6E-10	0.100	2.1E-10	1.1E-10	6.6E-11	4.0E-11	3.2E-11
		M	0.020	3.6E-10	0.010	2.8E-10	1.5E-10	9.1E-11	5.9E-11	4.7E-11
		S	0.020	3.7E-10	0.010	2.9E-10	1.5E-10	9.4E-11	6.1E-11	4.9E-11
Sb-117	2.80 h	F	0.200	7.7E-11	0.100	6.0E-11	2.9E-11	1.8E-11	1.0E-11	8.5E-12
		M	0.020	1.2E-10	0.010	9.1E-11	4.6E-11	3.0E-11	2.0E-11	1.6E-11
		S	0.020	1.3E-10	0.010	9.5E-11	4.8E-11	3.1E-11	2.2E-11	1.7E-11
Sb-118m	5.00 h	F	0.200	7.3E-10	0.100	6.2E-10	3.3E-10	2.0E-10	1.2E-10	9.3E-11
		M	0.020	9.3E-10	0.010	7.6E-10	4.0E-10	2.5E-10	1.5E-10	1.2E-10
		S	0.020	9.5E-10	0.010	7.8E-10	4.1E-10	2.5E-10	1.5E-10	1.2E-10
Sb-119	1.59 d	F	0.200	2.7E-10	0.100	2.0E-10	9.4E-11	5.5E-11	2.9E-11	2.3E-11
		M	0.020	4.0E-10	0.010	2.8E-10	1.3E-10	7.9E-11	4.4E-11	3.5E-11
		S	0.020	4.1E-10	0.010	2.9E-10	1.4E-10	8.2E-11	4.5E-11	3.6E-11
Sb-120	5.76 d	F	0.200	4.1E-09	0.100	3.3E-09	1.8E-09	1.1E-09	6.7E-10	5.5E-10
		M	0.020	6.3E-09	0.010	5.0E-09	2.8E-09	1.8E-09	1.3E-09	1.0E-09
		S	0.020	6.6E-09	0.010	5.3E-09	2.9E-09	1.9E-09	1.4E-09	1.1E-09

a   Dose coefficients for this element are based on age-specific biokinetic data

Table A.2.—(continued)

Nuclide	Physical half-life	Type	f_1 <1y	e(τ) 3 Months	f_1 ≥1y	e(t) 1 Year	5 Years	10 Years	15 Years	Adult
Sb-120	0.265 h	F	0.200	4.6E-11	0.100	3.1E-11	1.4E-11	8.9E-12	5.4E-12	4.6E-12
		M	0.020	6.6E-11	0.010	4.4E-11	2.0E-11	1.3E-11	8.3E-12	7.0E-12
		S	0.020	6.8E-11	0.010	4.6E-11	2.1E-11	1.4E-11	8.7E-12	7.3E-12
Sb-122	2.70 d	F	0.200	4.2E-09	0.100	2.8E-09	1.4E-09	8.4E-10	4.4E-10	3.6E-10
		M	0.020	8.3E-09	0.010	5.7E-09	2.8E-09	1.8E-09	1.3E-09	1.0E-09
		S	0.020	8.8E-09	0.010	6.1E-09	3.0E-09	2.0E-09	1.4E-09	1.1E-09
Sb-124	60.2 d	F	0.200	1.2E-08	0.100	8.8E-09	4.3E-09	2.6E-09	1.6E-09	1.3E-09
		M	0.020	3.1E-08	0.010	2.4E-08	1.4E-08	9.6E-09	7.7E-09	6.4E-09
		S	0.020	3.9E-08	0.010	3.1E-08	1.8E-08	1.3E-08	1.0E-08	8.6E-09
Sb-124m	0.337 h	F	0.200	2.7E-11	0.100	1.9E-11	9.0E-12	5.6E-12	3.4E-12	2.8E-12
		M	0.020	4.3E-11	0.010	3.1E-11	1.5E-11	9.6E-12	6.5E-12	5.4E-12
		S	0.020	4.6E-11	0.010	3.3E-11	1.6E-11	1.0E-11	7.2E-12	5.9E-12
Sb-125	2.77 y	F	0.200	8.7E-09	0.100	6.8E-09	3.7E-09	2.3E-09	1.5E-09	1.4E-09
		M	0.020	2.0E-08	0.010	1.6E-08	1.0E-08	6.8E-09	5.8E-09	4.8E-09
		S	0.020	4.2E-08	0.010	3.8E-08	2.4E-08	1.6E-08	1.4E-08	1.2E-08
Sb-126	12.4 d	F	0.200	8.8E-09	0.100	6.6E-09	3.3E-09	2.1E-09	1.2E-09	1.0E-09
		M	0.020	1.7E-08	0.010	1.3E-08	7.4E-09	5.1E-09	3.5E-09	2.8E-09
		S	0.020	1.9E-08	0.010	1.5E-08	8.2E-09	5.0E-09	4.0E-09	3.2E-09
Sb-126m	0.317 h	F	0.200	1.2E-10	0.100	8.2E-11	3.8E-11	2.4E-11	1.5E-11	1.2E-11
		M	0.020	1.7E-10	0.010	1.2E-10	5.5E-11	3.5E-11	2.3E-11	1.9E-11
		S	0.020	1.8E-10	0.010	1.2E-10	5.7E-11	3.7E-11	2.4E-11	2.0E-11
Sb-127	3.85 d	F	0.200	5.1E-09	0.100	3.5E-09	1.6E-09	9.7E-10	5.2E-10	4.3E-10
		M	0.020	1.0E-08	0.010	7.3E-09	3.9E-09	2.7E-09	2.1E-09	1.7E-09
		S	0.020	1.1E-08	0.010	7.9E-09	4.2E-09	3.0E-09	2.3E-09	1.9E-09
Sb-128	9.01 h	F	0.200	2.1E-09	0.100	1.7E-09	8.3E-10	5.1E-10	2.9E-10	2.3E-10
		M	0.020	3.3E-09	0.010	2.5E-09	1.2E-09	7.9E-10	5.0E-10	4.0E-10
		S	0.020	3.4E-09	0.010	2.6E-09	1.3E-09	8.3E-10	5.2E-10	4.2E-10
Sb-129	0.173 h	F	0.200	9.8E-11	0.100	6.9E-11	3.2E-11	2.0E-11	1.2E-11	1.0E-11
		M	0.020	1.3E-10	0.010	9.2E-11	4.3E-11	2.7E-11	1.7E-11	1.4E-11
		S	0.020	1.4E-10	0.010	9.4E-11	4.4E-11	2.8E-11	1.8E-11	1.5E-11
Sb-129	4.32 h	F	0.200	1.1E-09	0.100	8.2E-10	3.8E-10	2.3E-10	1.3E-10	1.0E-10
		M	0.020	2.0E-09	0.010	1.4E-09	6.8E-10	4.4E-10	2.9E-10	2.3E-10
		S	0.020	2.1E-09	0.010	1.5E-09	7.2E-10	4.6E-10	3.0E-10	2.5E-10
Sb-130	0.667 h	F	0.200	3.0E-10	0.100	2.2E-10	1.1E-10	6.6E-11	4.0E-11	3.3E-11
		M	0.020	4.5E-10	0.010	3.2E-10	1.6E-10	9.8E-11	6.3E-11	5.1E-11
		S	0.020	4.6E-10	0.010	3.3E-10	1.6E-10	1.0E-10	6.5E-11	5.3E-11
Sb-131	0.383 h	F	0.200	3.5E-10	0.100	2.8E-10	1.4E-10	7.7E-11	4.6E-11	3.5E-11
		M	0.020	3.9E-10	0.010	2.6E-10	1.3E-10	8.0E-11	5.3E-11	4.4E-11
		S	0.020	3.8E-10	0.010	2.6E-10	1.2E-10	7.9E-11	5.3E-11	4.4E-11
Tellurium[a]										
Te-116	2.49 h	F	0.600	5.3E-10	0.300	4.2E-10	2.1E-10	1.3E-10	7.2E-11	5.8E-11
		M	0.200	8.6E-10	0.100	6.4E-10	3.2E-10	2.0E-10	1.3E-10	1.0E-10
		S	0.020	9.1E-10	0.010	6.7E-10	3.3E-10	2.1E-10	1.4E-10	1.1E-10
Te-121	17.0 d	F	0.600	1.7E-09	0.300	1.4E-09	7.2E-10	4.6E-10	2.9E-10	2.4E-10
		M	0.200	2.3E-09	0.100	1.9E-09	1.0E-09	6.8E-10	4.7E-10	3.8E-10
		S	0.020	2.4E-09	0.010	2.0E-09	1.1E-09	7.2E-10	5.1E-10	4.1E-10
Te-121m	154 d	F	0.600	1.4E-08	0.300	1.0E-08	5.3E-09	3.3E-09	2.1E-09	1.8E-09
		M	0.200	1.9E-08	0.100	1.5E-08	8.8E-09	6.1E-09	5.1E-09	4.2E-09
		S	0.020	2.3E-08	0.010	1.9E-08	1.2E-08	8.1E-09	6.9E-09	5.7E-09

a Dose coefficients for this element are based on age-specific biokinetic data

Table A.2.—*(continued)*

Nuclide	Physical half-life	Type	f_1 <1y	$e(\tau)$ 3 Months	f_1 ≥1y	$e(\tau)$ 1 Year	5 Years	10 Years	15 Years	Adult
Te-123	1.00E+13 y	F	0.600	1.1E-08	0.300	9.1E-09	6.2E-09	4.8E-09	4.0E-09	3.9E-09
		M	0.200	5.6E-09	0.100	4.4E-09	3.0E-09	2.3E-09	2.0E-09	1.9E-09
		S	0.020	5.3E-09	0.010	5.0E-09	3.5E-09	2.4E-09	2.1E-09	2.0E-09
Te-123m	120 d	F	0.600	9.8E-09	0.300	6.8E-09	3.4E-09	1.9E-09	1.1E-09	9.5E-10
		M	0.200	1.8E-08	0.100	1.3E-08	8.0E-09	5.7E-09	5.0E-09	4.0E-09
		S	0.020	2.0E-08	0.010	1.6E-08	9.8E-09	7.1E-09	6.3E-09	5.1E-09
Te-125m	58.0 d	F	0.600	6.2E-09	0.300	4.2E-09	2.0E-09	1.1E-09	6.1E-10	5.1E-10
		M	0.200	1.5E-08	0.100	1.1E-08	6.6E-09	4.8E-09	4.3E-09	3.4E-09
		S	0.020	1.7E-08	0.010	1.3E-08	7.8E-09	5.8E-09	5.3E-09	4.2E-09
Te-127	9.35 h	F	0.600	4.3E-10	0.300	3.2E-10	1.4E-10	8.5E-11	4.5E-11	3.9E-11
		M	0.200	1.0E-09	0.100	7.3E-10	3.6E-10	2.4E-10	1.6E-10	1.3E-10
		S	0.020	1.2E-09	0.010	7.9E-10	3.9E-10	2.6E-10	1.7E-10	1.4E-10
Te-127m	109 d	F	0.600	2.1E-08	0.300	1.4E-08	6.5E-09	3.5E-09	2.0E-09	1.5E-09
		M	0.200	3.5E-08	0.100	2.6E-08	1.5E-08	1.1E-08	9.2E-09	7.4E-09
		S	0.020	4.1E-08	0.010	3.3E-08	2.0E-08	1.4E-08	1.2E-08	9.8E-09
Te-129	1.16 h	F	0.600	1.8E-10	0.300	1.2E-10	5.1E-11	3.2E-11	1.9E-11	1.6E-11
		M	0.200	3.3E-10	0.100	2.2E-10	9.9E-11	6.5E-11	4.4E-11	3.7E-11
		S	0.020	3.5E-10	0.010	2.3E-10	1.0E-10	6.9E-11	4.7E-11	3.9E-11
Te-129m	33.6 d	F	0.600	2.0E-08	0.300	1.3E-08	5.8E-09	3.1E-09	1.7E-09	1.3E-09
		M	0.200	3.5E-08	0.100	2.6E-08	1.4E-08	9.8E-09	8.0E-09	6.6E-09
		S	0.020	3.8E-08	0.010	2.9E-08	1.7E-08	1.2E-08	9.6E-09	7.9E-09
Te-131	0.417 h	F	0.600	2.3E-10	0.300	2.0E-10	9.9E-11	5.3E-11	3.3E-11	2.3E-11
		M	0.200	2.6E-10	0.100	1.7E-10	8.1E-11	5.2E-11	3.5E-11	2.8E-11
		S	0.020	2.4E-10	0.010	1.6E-10	7.4E-11	4.9E-11	3.3E-11	2.8E-11
Te-131m	1.25 d	F	0.600	8.7E-09	0.300	7.6E-09	3.9E-09	2.0E-09	1.2E-09	8.6E-10
		M	0.200	7.9E-09	0.100	5.8E-09	3.0E-09	1.9E-09	1.2E-09	9.4E-10
		S	0.020	7.0E-09	0.010	5.1E-09	2.6E-09	1.8E-09	1.1E-09	9.1E-10
Te-132	3.26 d	F	0.600	2.2E-08	0.300	1.8E-08	8.5E-09	4.2E-09	2.6E-09	1.8E-09
		M	0.200	1.6E-08	0.100	1.3E-08	6.4E-09	4.0E-09	2.6E-09	2.0E-09
		S	0.020	1.5E-08	0.010	1.1E-08	5.8E-09	3.8E-09	2.5E-09	2.0E-09
Te-133	0.207 h	F	0.600	2.4E-10	0.300	2.1E-10	9.6E-11	4.6E-11	2.8E-11	1.9E-11
		M	0.200	2.0E-10	0.100	1.3E-10	6.1E-11	3.8E-11	2.4E-11	2.0E-11
		S	0.020	1.7E-10	0.010	1.2E-10	5.4E-11	3.5E-11	2.2E-11	1.9E-11
Te-133m	0.923 h	F	0.600	1.0E-09	0.300	8.9E-10	4.1E-10	2.0E-10	1.2E-10	8.1E-11
		M	0.200	8.5E-10	0.100	5.8E-10	2.8E-10	1.7E-10	1.1E-10	8.7E-11
		S	0.020	7.4E-10	0.010	5.1E-10	2.5E-10	1.6E-10	1.0E-10	8.4E-11
Te-134	0.696 h	F	0.600	4.7E-10	0.300	3.7E-10	1.8E-10	1.0E-10	6.0E-11	4.7E-11
		M	0.200	5.5E-10	0.100	3.9E-10	1.9E-10	1.2E-10	8.1E-11	6.6E-11
		S	0.020	5.6E-10	0.010	4.0E-10	1.9E-10	1.3E-10	8.4E-11	6.8E-11
Iodine[a]										
I-120	1.35 h	F	1.000	1.3E-09	1.000	1.0E-09	4.8E-10	2.3E-10	1.4E-10	1.0E-10
		M	0.200	1.1E-09	0.100	7.3E-10	3.4E-10	2.1E-10	1.3E-10	1.0E-10
		S	0.020	1.0E-09	0.010	6.9E-10	3.2E-10	2.0E-10	1.2E-10	1.0E-10
I-120m	0.883 h	F	1.000	8.6E-10	1.000	6.9E-10	3.3E-10	1.8E-10	1.1E-10	8.2E-11
		M	0.200	8.2E-10	0.100	5.9E-10	2.9E-10	1.8E-10	1.1E-10	8.7E-11
		S	0.020	8.2E-10	0.010	5.8E-10	2.8E-10	1.8E-10	1.1E-10	8.8E-11
I-121	2.12 h	F	1.000	2.3E-10	1.000	2.1E-10	1.1E-10	6.0E-11	3.8E-11	2.7E-11
		M	0.200	2.1E-10	0.100	1.5E-10	7.8E-11	4.9E-11	3.2E-11	2.5E-11
		S	0.020	1.9E-10	0.010	1.4E-10	7.0E-11	4.5E-11	3.0E-11	2.4E-11

a Dose coefficients for this element are based on age-specific biokinetic data

Table A.2.—(continued)

Nuclide	Physical half-life	Type	f_1 <1y	e(τ) 3 Months	f_1 ≥1y	e(τ) 1 Year	5 Years	10 Years	15 Years	Adult
I-123	13.2 h	F	1.000	8.7E-10	1.000	7.9E-10	3.8E-10	1.8E-10	1.1E-10	7.4E-11
		M	0.200	5.3E-10	0.100	3.9E-10	2.0E-10	1.2E-10	8.2E-11	6.4E-11
		S	0.020	4.3E-10	0.010	3.2E-10	1.7E-10	1.1E-10	7.6E-11	6.0E-11
I-124	4.18 d	F	1.000	4.7E-08	1.000	4.5E-08	2.2E-08	1.1E-08	6.7E-09	4.4E-09
		M	0.200	1.4E-08	0.100	9.3E-09	4.6E-09	2.5E-09	1.6E-09	1.2E-09
		S	0.020	6.2E-09	0.010	4.4E-09	2.2E-09	1.4E-09	9.4E-10	7.7E-10
I-125	60.1 d	F	1.000	2.0E-08	1.000	2.3E-08	1.5E-08	1.1E-08	7.2E-09	5.1E-09
		M	0.200	6.9E-09	0.100	5.6E-09	3.6E-09	2.6E-09	1.8E-09	1.4E-09
		S	0.020	2.4E-09	0.010	1.8E-09	1.0E-09	6.7E-10	4.8E-10	3.8E-10
I-126	13.0 d	F	1.000	8.1E-08	1.000	8.3E-08	4.5E-08	2.4E-08	1.5E-08	9.8E-09
		M	0.200	2.4E-08	0.100	1.7E-08	9.5E-09	5.5E-09	3.8E-09	2.7E-09
		S	0.020	8.3E-09	0.010	5.9E-09	3.3E-09	2.2E-09	1.8E-09	1.4E-09
I-128	0.416 h	F	1.000	1.5E-10	1.000	1.1E-10	4.7E-11	2.7E-11	1.6E-11	1.3E-11
		M	0.200	1.9E-10	0.100	1.2E-10	5.3E-11	3.4E-11	2.2E-11	1.9E-11
		S	0.020	1.9E-10	0.010	1.2E-10	5.4E-11	3.5E-11	2.3E-11	2.0E-11
I-129	1.57E+07 y	F	1.000	7.2E-08	1.000	8.6E-08	6.1E-08	6.7E-08	4.6E-08	3.6E-08
		M	0.200	3.6E-08	0.100	3.3E-08	2.4E-08	2.4E-08	1.9E-08	1.5E-08
		S	0.020	2.9E-08	0.010	2.6E-08	1.8E-08	1.3E-08	1.1E-08	9.8E-09
I-130	12.4 h	F	1.000	8.2E-09	1.000	7.4E-09	3.5E-09	1.6E-09	1.0E-09	6.7E-10
		M	0.200	4.3E-09	0.100	3.1E-09	1.5E-09	9.2E-10	5.8E-10	4.5E-10
		S	0.020	3.3E-09	0.010	2.4E-09	1.2E-09	7.9E-10	5.1E-10	4.1E-10
I-131	8.04 d	F	1.000	7.2E-08	1.000	7.2E-08	3.7E-08	1.9E-08	1.1E-08	7.4E-09
		M	0.200	2.2E-08	0.100	1.5E-08	8.2E-09	4.7E-09	3.4E-09	2.4E-09
		S	0.020	8.8E-09	0.010	6.2E-09	3.5E-09	2.4E-09	2.0E-09	1.6E-09
I-132	2.30 h	F	1.000	1.1E-09	1.000	9.6E-10	4.5E-10	2.2E-10	1.3E-10	9.4E-11
		M	0.200	9.9E-10	0.100	7.3E-10	3.6E-10	2.2E-10	1.4E-10	1.1E-10
		S	0.020	9.3E-10	0.010	6.8E-10	3.4E-10	2.1E-10	1.4E-10	1.1E-10
I-132m	1.39 h	F	1.000	9.6E-10	1.000	8.4E-10	4.0E-10	1.9E-10	1.2E-10	7.9E-11
		M	0.200	7.2E-10	0.100	5.3E-10	2.6E-10	1.6E-10	1.1E-10	8.7E-11
		S	0.020	6.6E-10	0.010	4.8E-10	2.4E-10	1.6E-10	1.1E-10	8.5E-11
I-133	20.8 h	F	1.000	1.9E-08	1.000	1.8E-08	8.3E-09	3.8E-09	2.2E-09	1.5E-09
		M	0.200	6.6E-09	0.100	4.4E-09	2.1E-09	1.2E-09	7.4E-10	5.5E-10
		S	0.020	3.8E-09	0.010	2.9E-09	1.4E-09	9.0E-10	5.3E-10	4.3E-10
I-134	0.876 h	F	1.000	4.6E-10	1.000	3.7E-10	1.8E-10	9.7E-11	5.9E-11	4.5E-11
		M	0.200	4.8E-10	0.100	3.4E-10	1.7E-10	1.0E-10	6.7E-11	5.4E-11
		S	0.020	4.8E-10	0.010	3.4E-10	1.7E-10	1.1E-10	6.8E-11	5.5E-11
I-135	6.61 h	F	1.000	4.1E-09	1.000	3.7E-09	1.7E-09	7.9E-10	4.8E-10	3.2E-10
		M	0.200	2.2E-09	0.100	1.6E-09	7.8E-10	4.7E-10	3.0E-10	2.4E-10
		S	0.020	1.8E-09	0.010	1.3E-09	6.5E-10	4.2E-10	2.7E-10	2.2E-10

Caesium[a]

Nuclide	Physical half-life	Type	f_1 <1y	e(τ) 3 Months	f_1 ≥1y	e(τ) 1 Year	5 Years	10 Years	15 Years	Adult
Cs-125	0.750 h	F	1.000	1.2E-10	1.000	8.3E-11	3.9E-11	2.4E-11	1.4E-11	1.2E-11
		M	0.200	2.0E-10	0.100	1.4E-10	6.5E-11	4.2E-11	2.7E-11	2.2E-11
		S	0.020	2.1E-10	0.010	1.4E-10	6.8E-11	4.4E-11	2.8E-11	2.3E-11
Cs-127	6.25 h	F	1.000	1.6E-10	1.000	1.3E-10	6.9E-11	4.2E-11	2.5E-11	2.0E-11
		M	0.200	2.8E-10	0.100	2.2E-10	1.1E-10	7.3E-11	4.6E-11	3.6E-11
		S	0.020	3.0E-10	0.010	2.3E-10	1.2E-10	7.6E-11	4.8E-11	3.8E-11
Cs-129	1.34 d	F	1.000	3.4E-10	1.000	2.8E-10	1.4E-10	8.7E-11	5.2E-11	4.2E-11
		M	0.200	5.7E-10	0.100	4.6E-10	2.4E-10	1.5E-10	9.1E-11	7.3E-11
		S	0.020	6.3E-10	0.010	4.9E-10	2.5E-10	1.6E-10	9.7E-11	7.7E-11

a   Dose coefficients for this element are based on age-specific biokinetic data

Table A.2.—(continued)

Nuclide	Physical half-life	Type	f_1 <1y	$e(\tau)$ 3 Months	f_1 ≥1y	$e(\tau)$ 1 Year	5 Years	10 Years	15 Years	Adult
Cs-130	0.498 h	F	1.000	8.3E-11	1.000	5.6E-11	2.5E-11	1.6E-11	9.4E-12	7.8E-12
		M	0.200	1.3E-10	0.100	8.7E-11	4.0E-11	2.5E-11	1.6E-11	1.4E-11
		S	0.020	1.4E-10	0.010	9.0E-11	4.1E-11	2.6E-11	1.7E-11	1.4E-11
Cs-131	9.69 d	F	1.000	2.4E-10	1.000	1.7E-10	8.4E-11	5.3E-11	3.2E-11	2.7E-11
		M	0.200	3.5E-10	0.100	2.6E-10	1.4E-10	8.5E-11	5.5E-11	4.4E-11
		S	0.020	3.8E-10	0.010	2.8E-10	1.4E-10	9.1E-11	5.9E-11	4.7E-11
Cs-132	6.48 d	F	1.000	1.5E-09	1.000	1.2E-09	6.4E-10	4.1E-10	2.7E-10	2.3E-10
		M	0.200	1.9E-09	0.100	1.5E-09	8.4E-10	5.4E-10	3.7E-10	2.9E-10
		S	0.020	2.0E-09	0.010	1.6E-09	8.7E-10	5.6E-10	3.8E-10	3.0E-10
Cs-134	2.06 y	F	1.000	1.1E-08	1.000	7.3E-09	5.2E-09	5.3E-09	6.3E-09	6.6E-09
		M	0.200	3.2E-08	0.100	2.6E-08	1.6E-08	1.2E-08	1.1E-08	9.1E-09
		S	0.020	7.0E-08	0.010	6.3E-08	4.1E-08	2.8E-08	2.3E-08	2.0E-08
Cs-134m	2.90 h	F	1.000	1.3E-10	1.000	8.6E-11	3.8E-11	2.5E-11	1.6E-11	1.4E-11
		M	0.200	3.3E-10	0.100	2.3E-10	1.2E-10	8.3E-11	6.6E-11	5.4E-11
		S	0.020	3.6E-10	0.010	2.5E-10	1.3E-10	9.2E-11	7.4E-11	6.0E-11
Cs-135	2.30E+06 y	F	1.000	1.7E-09	1.000	9.9E-10	6.2E-10	6.1E-10	6.8E-10	6.9E-10
		M	0.200	1.2E-08	0.100	9.3E-09	5.7E-09	4.1E-09	3.8E-09	3.1E-09
		S	0.020	2.7E-08	0.010	2.4E-08	1.6E-08	1.1E-08	9.5E-09	8.6E-09
Cs-135m	0.883 h	F	1.000	9.2E-11	1.000	7.8E-11	4.1E-11	2.4E-11	1.5E-11	1.2E-11
		M	0.200	1.2E-10	0.100	9.9E-11	5.2E-11	3.2E-11	1.9E-11	1.5E-11
		S	0.020	1.2E-10	0.010	1.0E-10	5.3E-11	3.3E-11	2.0E-11	1.6E-11
Cs-136	13.1 d	F	1.000	7.3E-09	1.000	5.2E-09	2.9E-09	2.0E-09	1.4E-09	1.2E-09
		M	0.200	1.3E-08	0.100	1.0E-08	6.0E-09	3.7E-09	3.1E-09	2.5E-09
		S	0.020	1.5E-08	0.010	1.1E-08	5.7E-09	4.1E-09	3.5E-09	2.8E-09
Cs-137	30.0 y	F	1.000	8.8E-09	1.000	5.4E-09	3.6E-09	3.7E-09	4.4E-09	4.6E-09
		M	0.200	3.6E-08	0.100	2.9E-08	1.8E-08	1.3E-08	1.1E-08	9.7E-09
		S	0.020	1.1E-07	0.010	1.0E-07	7.0E-08	4.8E-08	4.2E-08	3.9E-08
Cs-138	0.536 h	F	1.000	2.6E-10	1.000	1.8E-10	8.1E-11	5.0E-11	2.9E-11	2.4E-11
		M	0.200	4.0E-10	0.100	2.7E-10	1.3E-10	7.8E-11	4.9E-11	4.1E-11
		S	0.020	4.2E-10	0.010	2.8E-10	1.3E-10	8.2E-11	5.1E-11	4.3E-11
Barium[a,b]										
Ba-126	1.61 h	F	0.600	6.7E-10	0.200	5.2E-10	2.4E-10	1.4E-10	6.9E-11	7.4E-11
		M	0.200	1.0E-09	0.100	7.0E-10	3.2E-10	2.0E-10	1.2E-10	1.0E-10
		S	0.020	1.1E-09	0.010	7.2E-10	3.3E-10	2.1E-10	1.3E-10	1.1E-10
Ba-128	2.43 d	F	0.600	5.9E-09	0.200	5.4E-09	2.5E-09	1.4E-09	7.4E-10	7.6E-10
		M	0.200	1.1E-08	0.100	7.8E-09	3.7E-09	2.4E-09	1.5E-09	1.3E-09
		S	0.020	1.2E-08	0.010	8.3E-09	4.0E-09	2.6E-09	1.6E-09	1.4E-09
Ba-131	11.8 d	F	0.600	2.1E-09	0.200	1.4E-09	7.1E-10	4.7E-10	3.1E-10	2.2E-10
		M	0.200	3.7E-09	0.100	3.1E-09	1.6E-09	1.1E-09	9.7E-10	7.6E-10
		S	0.020	4.0E-09	0.010	3.0E-09	1.8E-09	1.3E-09	1.1E-09	8.7E-10
Ba-131m	0.243 h	F	0.600	2.7E-11	0.200	2.1E-11	1.0E-11	6.7E-12	4.7E-12	4.0E-12
		M	0.200	4.8E-11	0.100	3.3E-11	1.7E-11	1.2E-11	9.0E-12	7.4E-12
		S	0.020	5.0E-11	0.010	3.5E-11	1.8E-11	1.2E-11	9.5E-12	7.8E-12
Ba-133	10.7 y	F	0.600	1.1E-08	0.200	4.5E-09	2.6E-09	3.7E-09	6.0E-09	1.5E-09
		M	0.200	1.5E-08	0.100	1.0E-08	6.4E-09	5.1E-09	5.5E-09	3.1E-09
		S	0.020	3.2E-08	0.010	2.9E-08	2.0E-08	1.3E-08	1.1E-08	1.0E-08
Ba-133m	1.62 d	F	0.600	1.4E-09	0.200	1.1E-09	4.9E-10	3.1E-10	1.5E-10	1.8E-10
		M	0.200	3.0E-09	0.100	2.2E-09	1.0E-09	6.9E-10	5.2E-10	4.2E-10
		S	0.020	3.1E-09	0.010	2.4E-09	1.1E-09	7.6E-10	5.8E-10	4.6E-10

a Dose coefficients for this element are based on age-specific biokinetic data

b The f_1 value for 1 to 15 year olds for Type F is 0.3

Table A.2.—(continued)

Nuclide	Physical half-life	Type	f_1 <1y	$e(\tau)$ 3 Months	f_1 ≥1y	$e(\tau)$ 1 Year	5 Years	10 Years	15 Years	Adult
Ba-135m	1.20 d	F	0.600	1.1E-09	0.200	1.0E-09	4.6E-10	2.5E-10	1.2E-10	1.4E-10
		M	0.200	2.4E-09	0.100	1.8E-09	8.9E-10	5.4E-10	4.1E-10	3.3E-10
		S	0.020	2.7E-09	0.010	1.9E-09	8.6E-10	5.9E-10	4.5E-10	3.6E-10
Ba-139	1.38 h	F	0.600	3.3E-10	0.200	2.4E-10	1.1E-10	6.0E-11	3.1E-11	3.4E-11
		M	0.200	5.4E-10	0.100	3.5E-10	1.6E-10	1.0E-10	6.6E-11	5.6E-11
		S	0.020	5.7E-10	0.010	3.6E-10	1.6E-10	1.1E-10	7.0E-11	5.9E-11
Ba-140	12.7 d	F	0.600	1.4E-08	0.200	7.8E-09	3.6E-09	2.4E-09	1.6E-09	1.0E-09
		M	0.200	2.7E-08	0.100	2.0E-08	1.1E-08	7.6E-09	6.2E-09	5.1E-09
		S	0.020	2.9E-08	0.010	2.2E-08	1.2E-08	8.6E-09	7.1E-09	5.8E-09
Ba-141	0.305 h	F	0.600	1.9E-10	0.200	1.4E-10	6.4E-11	3.8E-11	2.1E-11	2.1E-11
		M	0.200	3.0E-10	0.100	2.0E-10	9.3E-11	5.9E-11	3.8E-11	3.2E-11
		S	0.020	3.2E-10	0.010	2.1E-10	9.7E-11	6.2E-11	4.0E-11	3.4E-11
Ba-142	0.177 h	F	0.600	1.3E-10	0.200	9.6E-11	4.5E-11	2.7E-11	1.6E-11	1.5E-11
		M	0.200	1.8E-10	0.100	1.3E-10	6.1E-11	3.9E-11	2.5E-11	2.1E-11
		S	0.020	1.9E-10	0.010	1.3E-10	6.2E-11	4.0E-11	2.6E-11	2.2E-11

Lanthanum

Nuclide	Physical half-life	Type	f_1 <1y	$e(\tau)$ 3 Months	f_1 ≥1y	$e(\tau)$ 1 Year	5 Years	10 Years	15 Years	Adult
La-131	0.983 h	F	0.005	1.2E-10	5.0E-04	8.7E-11	4.2E-11	2.6E-11	1.5E-11	1.3E-11
		M	0.005	1.8E-10	5.0E-04	1.3E-10	6.4E-11	4.1E-11	2.8E-11	2.3E-11
La-132	4.80 h	F	0.005	1.0E-09	5.0E-04	7.7E-10	3.7E-10	2.2E-10	1.2E-10	1.0E-10
		M	0.005	1.5E-09	5.0E-04	1.1E-09	5.4E-10	3.4E-10	2.0E-10	1.6E-10
La-135	19.5 h	F	0.005	1.0E-10	5.0E-04	7.7E-11	3.8E-11	2.3E-11	1.3E-11	1.0E-11
		M	0.005	1.3E-10	5.0E-04	1.0E-10	4.9E-11	3.0E-11	1.7E-11	1.4E-11
La-137	6.00E+04 y	F	0.005	2.5E-08	5.0E-04	2.3E-08	1.5E-08	1.1E-08	8.9E-09	8.7E-09
		M	0.005	8.6E-09	5.0E-04	8.1E-09	5.6E-09	4.0E-09	3.6E-09	3.6E-09
La-138	1.35E+11 y	F	0.005	3.7E-07	5.0E-04	3.5E-07	2.4E-07	1.8E-07	1.6E-07	1.5E-07
		M	0.005	1.3E-07	5.0E-04	1.2E-07	9.1E-08	6.8E-08	6.4E-08	6.4E-08
La-140	1.68 d	F	0.005	5.8E-09	5.0E-04	4.2E-09	2.0E-09	1.2E-09	6.9E-10	5.7E-10
		M	0.005	8.8E-09	5.0E-04	6.3E-09	3.1E-09	2.0E-09	1.3E-09	1.1E-09
La-141	3.93 h	F	0.005	8.6E-10	5.0E-04	5.5E-10	2.3E-10	1.4E-10	7.5E-11	6.3E-11
		M	0.005	1.4E-09	5.0E-04	9.3E-10	4.3E-10	2.8E-10	1.8E-10	1.5E-10
La-142	1.54 h	F	0.005	5.3E-10	5.0E-04	3.8E-10	1.8E-10	1.1E-10	6.3E-11	5.2E-11
		M	0.005	8.1E-10	5.0E-04	5.7E-10	2.7E-10	1.7E-10	1.1E-10	8.9E-11
La-143	0.237 h	F	0.005	1.4E-10	5.0E-04	8.6E-11	3.7E-11	2.3E-11	1.4E-11	1.2E-11
		M	0.005	2.1E-10	5.0E-04	1.3E-10	6.0E-11	3.9E-11	2.5E-11	2.1E-11

Cerium[a]

Nuclide	Physical half-life	Type	f_1 <1y	$e(\tau)$ 3 Months	f_1 ≥1y	$e(\tau)$ 1 Year	5 Years	10 Years	15 Years	Adult
Ce-134	3.00 d	F	0.005	7.6E-09	5.0E-04	5.3E-09	2.3E-09	1.4E-09	7.7E-10	5.7E-10
		M	0.005	1.1E-08	5.0E-04	7.6E-09	3.7E-09	2.4E-09	1.5E-09	1.3E-09
		S	0.005	1.2E-08	5.0E-04	8.0E-09	3.8E-09	2.5E-09	1.6E-09	1.3E-09
Ce-135	17.6 h	F	0.005	2.3E-09	5.0E-04	1.7E-09	8.5E-10	5.3E-10	3.0E-10	2.4E-10
		M	0.005	3.6E-09	5.0E-04	2.7E-09	1.4E-09	8.9E-10	5.9E-10	4.8E-10
		S	0.005	3.7E-09	5.0E-04	2.8E-09	1.4E-09	9.4E-10	6.3E-10	5.0E-10
Ce-137	9.00 h	F	0.005	7.5E-11	5.0E-04	5.6E-11	2.7E-11	1.6E-11	8.7E-12	7.0E-12
		M	0.005	1.1E-10	5.0E-04	7.6E-11	3.6E-11	2.2E-11	1.2E-11	9.8E-12
		S	0.005	1.1E-10	5.0E-04	7.8E-11	3.7E-11	2.3E-11	1.3E-11	1.0E-11
Ce-137m	1.43 d	F	0.005	1.6E-09	5.0E-04	1.1E-09	4.6E-10	2.8E-10	1.5E-10	1.2E-10
		M	0.005	3.1E-09	5.0E-04	2.2E-09	1.1E-09	6.7E-10	5.1E-10	4.1E-10
		S	0.005	3.3E-09	5.0E-04	2.3E-09	1.0E-09	7.3E-10	5.6E-10	4.4E-10

a Dose coefficients for this element are based on age-specific biokinetic data

Table A.2.—(continued)

Nuclide	Physical half-life	Type	f_1 <1y	$e(\tau)$ 3 Months	f_1 ≥1y	$e(\tau)$ 1 Year	5 Years	10 Years	15 Years	Adult
Ce-139	138 d	F	0.005	1.1E-08	5.0E-04	8.5E-09	4.5E-09	2.8E-09	1.8E-09	1.5E-09
		M	0.005	7.5E-09	5.0E-04	6.1E-09	3.6E-09	2.5E-09	2.1E-09	1.7E-09
		S	0.005	7.8E-09	5.0E-04	6.3E-09	3.9E-09	2.7E-09	2.4E-09	1.9E-09
Ce-141	32.5 d	F	0.005	1.1E-08	5.0E-04	7.3E-09	3.5E-09	2.0E-09	1.2E-09	9.3E-10
		M	0.005	1.4E-08	5.0E-04	1.1E-08	6.3E-09	4.6E-09	4.1E-09	3.2E-09
		S	0.005	1.6E-08	5.0E-04	1.2E-08	7.1E-09	5.3E-09	4.8E-09	3.8E-09
Ce-143	1.38 d	F	0.005	3.6E-09	5.0E-04	2.3E-09	1.0E-09	6.2E-10	3.3E-10	2.7E-10
		M	0.005	5.6E-09	5.0E-04	3.9E-09	1.9E-09	1.3E-09	9.3E-10	7.5E-10
		S	0.005	5.9E-09	5.0E-04	4.1E-09	2.1E-09	1.4E-09	1.0E-09	8.3E-10
Ce-144	284 d	F	0.005	3.6E-07	5.0E-04	2.7E-07	1.4E-07	7.8E-08	4.8E-08	4.0E-08
		M	0.005	1.9E-07	5.0E-04	1.6E-07	8.8E-08	5.5E-08	4.1E-08	3.6E-08
		S	0.005	2.1E-07	5.0E-04	1.8E-07	1.1E-07	7.3E-08	5.8E-08	5.3E-08

Praseodymium

Nuclide	Physical half-life	Type	f_1 <1y	$e(\tau)$ 3 Months	f_1 ≥1y	$e(\tau)$ 1 Year	5 Years	10 Years	15 Years	Adult
Pr-136	0.218 h	M	0.005	1.3E-10	5.0E-04	8.8E-11	4.2E-11	2.6E-11	1.6E-11	1.3E-11
		S	0.005	1.3E-10	5.0E-04	9.0E-11	4.3E-11	2.7E-11	1.7E-11	1.4E-11
Pr-137	1.28 h	M	0.005	1.8E-10	5.0E-04	1.3E-10	6.1E-11	3.9E-11	2.4E-11	2.0E-11
		S	0.005	1.9E-10	5.0E-04	1.3E-10	6.4E-11	4.0E-11	2.5E-11	2.1E-11
Pr-138m	2.10 h	M	0.005	5.9E-10	5.0E-04	4.5E-10	2.3E-10	1.4E-10	9.0E-11	7.2E-11
		S	0.005	6.0E-10	5.0E-04	4.7E-10	2.4E-10	1.5E-10	9.3E-11	7.4E-11
Pr-139	4.51 h	M	0.005	1.5E-10	5.0E-04	1.1E-10	5.5E-11	3.5E-11	2.3E-11	1.8E-11
		S	0.005	1.6E-10	5.0E-04	1.2E-10	5.7E-11	3.7E-11	2.4E-11	2.0E-11
Pr-142	19.1 h	M	0.005	5.3E-09	5.0E-04	3.5E-09	1.6E-09	1.0E-09	6.2E-10	5.2E-10
		S	0.005	5.5E-09	5.0E-04	3.7E-09	1.7E-09	1.1E-09	6.6E-10	5.5E-10
Pr-142m	0.243 h	M	0.005	6.7E-11	5.0E-04	4.5E-11	2.0E-11	1.3E-11	7.9E-12	6.6E-12
		S	0.005	7.0E-11	5.0E-04	4.7E-11	2.2E-11	1.4E-11	8.4E-12	7.0E-12
Pr-143	13.6 d	M	0.005	1.2E-08	5.0E-04	8.4E-09	4.6E-09	3.2E-09	2.7E-09	2.2E-09
		S	0.005	1.3E-08	5.0E-04	9.2E-09	5.1E-09	3.6E-09	3.0E-09	2.4E-09
Pr-144	0.288 h	M	0.005	1.9E-10	5.0E-04	1.2E-10	5.0E-11	3.2E-11	2.1E-11	1.8E-11
		S	0.005	1.9E-10	5.0E-04	1.2E-10	5.2E-11	3.4E-11	2.1E-11	1.8E-11
Pr-145	5.98 h	M	0.005	1.6E-09	5.0E-04	1.0E-09	4.7E-10	3.0E-10	1.9E-10	1.6E-10
		S	0.005	1.6E-09	5.0E-04	1.1E-09	4.9E-10	3.2E-10	2.0E-10	1.7E-10
Pr-147	0.227 h	M	0.005	1.5E-10	5.0E-04	1.0E-10	4.8E-11	3.1E-11	2.1E-11	1.8E-11
		S	0.005	1.6E-10	5.0E-04	1.1E-10	5.0E-11	3.3E-11	2.2E-11	1.8E-11

Neodymium

Nuclide	Physical half-life	Type	f_1 <1y	$e(\tau)$ 3 Months	f_1 ≥1y	$e(\tau)$ 1 Year	5 Years	10 Years	15 Years	Adult
Nd-136	0.844 h	M	0.005	4.6E-10	5.0E-04	3.2E-10	1.6E-10	9.8E-11	6.3E-11	5.1E-11
		S	0.005	4.8E-10	5.0E-04	3.3E-10	1.6E-10	1.0E-10	6.6E-11	5.4E-11
Nd-138	5.04 h	M	0.005	2.3E-09	5.0E-04	1.7E-09	7.7E-10	4.8E-10	2.8E-10	2.3E-10
		S	0.005	2.4E-09	5.0E-04	1.8E-09	8.0E-10	5.0E-10	3.0E-10	2.5E-10
Nd-139	0.495 h	M	0.005	9.0E-11	5.0E-04	6.2E-11	3.0E-11	1.9E-11	1.2E-11	9.9E-12
		S	0.005	9.4E-11	5.0E-04	6.4E-11	3.1E-11	2.0E-11	1.3E-11	1.0E-11
Nd-139m	5.50 h	M	0.005	1.1E-09	5.0E-04	8.8E-10	4.5E-10	2.9E-10	1.8E-10	1.5E-10
		S	0.005	1.2E-09	5.0E-04	9.1E-10	4.6E-10	3.0E-10	1.9E-10	1.5E-10
Nd-141	2.49 h	M	0.005	4.1E-11	5.0E-04	3.1E-11	1.5E-11	9.6E-12	6.0E-12	4.8E-12
		S	0.005	4.3E-11	5.0E-04	3.2E-11	1.6E-11	1.0E-11	6.2E-12	5.0E-12
Nd-147	11.0 d	M	0.005	1.1E-08	5.0E-04	8.0E-09	4.5E-09	3.2E-09	2.6E-09	2.1E-09
		S	0.005	1.2E-08	5.0E-04	8.6E-09	4.9E-09	3.5E-09	3.0E-09	2.4E-09

Table A.2.—(*continued*)

Nuclide	Physical half-life	Type	f_1 <1y	$e(\tau)$ 3 Months	f_1 ≥1y	$e(\tau)$ 1 Year	5 Years	10 Years	15 Years	Adult
Nd-149	1.73 h	M	0.005	6.8E-10	5.0E-04	4.6E-10	2.2E-10	1.5E-10	1.0E-10	8.4E-11
		S	0.005	7.1E-10	5.0E-04	4.8E-10	2.3E-10	1.5E-10	1.1E-10	8.9E-11
Nd-151	0.207 h	M	0.005	1.5E-10	5.0E-04	9.9E-11	4.6E-11	3.0E-11	2.0E-11	1.7E-11
		S	0.005	1.5E-10	5.0E-04	1.0E-10	4.8E-11	3.1E-11	2.1E-11	1.7E-11
Promethium										
Pm-141	0.348 h	M	0.005	1.4E-10	5.0E-04	9.4E-11	4.3E-11	2.7E-11	1.7E-11	1.4E-11
		S	0.005	1.5E-10	5.0E-04	9.7E-11	4.4E-11	2.8E-11	1.8E-11	1.5E-11
Pm-143	265 d	M	0.005	6.2E-09	5.0E-04	5.4E-09	3.3E-09	2.2E-09	1.7E-09	1.5E-09
		S	0.005	5.5E-09	5.0E-04	4.8E-09	3.1E-09	2.1E-09	1.7E-09	1.4E-09
Pm-144	363 d	M	0.005	3.1E-08	5.0E-04	2.8E-08	1.8E-08	1.2E-08	9.3E-09	8.2E-09
		S	0.005	2.6E-08	5.0E-04	2.4E-08	1.6E-08	1.1E-08	8.9E-09	7.5E-09
Pm-145	17.7 y	M	0.005	1.1E-08	5.0E-04	9.8E-09	6.4E-09	4.3E-09	3.7E-09	3.6E-09
		S	0.005	7.1E-09	5.0E-04	6.5E-09	4.3E-09	2.9E-09	2.4E-09	2.3E-09
Pm-146	5.53 y	M	0.005	6.4E-08	5.0E-04	5.9E-08	3.9E-08	2.6E-08	2.2E-08	2.1E-08
		S	0.005	5.3E-08	5.0E-04	4.9E-08	3.3E-08	2.2E-08	1.9E-08	1.7E-08
Pm-147	2.62 y	M	0.005	2.1E-08	5.0E-04	1.8E-08	1.1E-08	7.0E-09	5.7E-09	5.0E-09
		S	0.005	1.9E-08	5.0E-04	1.6E-08	1.0E-08	6.8E-09	5.8E-09	4.9E-09
Pm-148	5.37 d	M	0.005	1.5E-08	5.0E-04	1.0E-08	5.2E-09	3.4E-09	2.4E-09	2.0E-09
		S	0.005	1.5E-08	5.0E-04	1.1E-08	5.5E-09	3.7E-09	2.6E-09	2.2E-09
Pm-148m	41.3 d	M	0.005	2.4E-08	5.0E-04	1.9E-08	1.1E-08	7.7E-09	6.3E-09	5.1E-09
		S	0.005	2.5E-08	5.0E-04	2.0E-08	1.2E-08	8.3E-09	7.1E-09	5.7E-09
Pm-149	2.21 d	M	0.005	5.0E-09	5.0E-04	3.5E-09	1.7E-09	1.1E-09	8.3E-10	6.7E-10
		S	0.005	5.3E-09	5.0E-04	3.6E-09	1.8E-09	1.2E-09	9.0E-10	7.3E-10
Pm-150	2.68 h	M	0.005	1.2E-09	5.0E-04	7.9E-10	3.8E-10	2.4E-10	1.5E-10	1.2E-10
		S	0.005	1.2E-09	5.0E-04	8.2E-10	3.9E-10	2.5E-10	1.6E-10	1.3E-10
Pm-151	1.18 d	M	0.005	3.3E-09	5.0E-04	2.5E-09	1.2E-09	8.3E-10	5.3E-10	4.3E-10
		S	0.005	3.4E-09	5.0E-04	2.6E-09	1.3E-09	7.9E-10	5.7E-10	4.6E-10
Samarium										
Sm-141	0.170 h	M	0.005	1.5E-10	5.0E-04	1.0E-10	4.7E-11	2.9E-11	1.8E-11	1.5E-11
Sm-141m	0.377 h	M	0.005	3.0E-10	5.0E-04	2.1E-10	9.7E-11	6.1E-11	3.9E-11	3.2E-11
Sm-142	1.21 h	M	0.005	7.5E-10	5.0E-04	4.8E-10	2.2E-10	1.4E-10	8.5E-11	7.1E-11
Sm-145	340 d	M	0.005	8.1E-09	5.0E-04	6.8E-09	4.0E-09	2.5E-09	1.9E-09	1.6E-09
Sm-146	1.03E+08 y	M	0.005	2.7E-05	5.0E-04	2.6E-05	1.7E-05	1.2E-05	1.1E-05	1.1E-05
Sm-147	1.06E+11 y	M	0.005	2.5E-05	5.0E-04	2.3E-05	1.6E-05	1.1E-05	9.6E-06	9.6E-06
Sm-151	90.0 y	M	0.005	1.1E-08	5.0E-04	1.0E-08	6.7E-09	4.5E-09	4.0E-09	4.0E-09
Sm-153	1.95 d	M	0.005	4.2E-09	5.0E-04	2.9E-09	1.5E-09	1.0E-09	7.9E-10	6.3E-10
Sm-155	0.368 h	M	0.005	1.5E-10	5.0E-04	9.9E-11	4.4E-11	2.9E-11	2.0E-11	1.7E-11
Sm-156	9.40 h	M	0.005	1.6E-09	5.0E-04	1.1E-09	5.8E-10	3.5E-10	2.7E-10	2.2E-10
Europium										
Eu-145	5.94 d	M	0.005	3.6E-09	5.0E-04	2.9E-09	1.6E-09	1.0E-09	6.8E-10	5.5E-10

Table A.2.—(continued)

Nuclide	Physical half-life	Type	f_1 <1y	$e(\tau)$ 3 Months	f_1 ≥1y	$e(\tau)$ 1 Year	5 Years	10 Years	15 Years	Adult
Eu-146	4.61 d	M	0.005	5.5E-09	5.0E-04	4.4E-09	2.4E-09	1.5E-09	1.0E-09	8.0E-10
Eu-147	24.0 d	M	0.005	4.9E-09	5.0E-04	3.7E-09	2.2E-09	1.6E-09	1.3E-09	1.1E-09
Eu-148	54.5 d	M	0.005	1.4E-08	5.0E-04	1.2E-08	6.8E-09	4.6E-09	3.2E-09	2.6E-09
Eu-149	93.1 d	M	0.005	1.6E-09	5.0E-04	1.3E-09	7.3E-10	4.7E-10	3.5E-10	2.9E-10
Eu-150	34.2 y	M	0.005	1.1E-07	5.0E-04	1.1E-07	7.8E-08	5.7E-08	5.3E-08	5.3E-08
Eu-150	12.6 h	M	0.005	1.6E-09	5.0E-04	1.1E-09	5.2E-10	3.4E-10	2.3E-10	1.9E-10
Eu-152	13.3 y	M	0.005	1.1E-07	5.0E-04	1.0E-07	7.0E-08	4.9E-08	4.3E-08	4.2E-08
Eu-152m	9.32 h	M	0.005	1.9E-09	5.0E-04	1.3E-09	6.6E-10	4.2E-10	2.4E-10	2.2E-10
Eu-154	8.80 y	M	0.005	1.6E-07	5.0E-04	1.5E-07	9.7E-08	6.5E-08	5.6E-08	5.3E-08
Eu-155	4.96 y	M	0.005	2.6E-08	5.0E-04	2.3E-08	1.4E-08	9.2E-09	7.6E-09	6.9E-09
Eu-156	15.2 d	M	0.005	1.9E-08	5.0E-04	1.4E-08	7.7E-09	5.3E-09	4.2E-09	3.4E-09
Eu-157	15.1 h	M	0.005	2.5E-09	5.0E-04	1.9E-09	8.9E-10	5.9E-10	3.5E-10	2.8E-10
Eu-158	0.765 h	M	0.005	4.3E-10	5.0E-04	2.9E-10	1.3E-10	8.5E-11	5.6E-11	4.7E-11
Gadolinium										
Gd-145	0.382 h	F	0.005	1.3E-10	5.0E-04	9.6E-11	4.7E-11	2.9E-11	1.7E-11	1.4E-11
		M	0.005	1.8E-10	5.0E-04	1.3E-10	6.2E-11	3.9E-11	2.4E-11	2.0E-11
Gd-146	48.3 d	F	0.005	2.9E-08	5.0E-04	2.3E-08	1.2E-08	7.8E-09	5.1E-09	4.4E-09
		M	0.005	2.8E-08	5.0E-04	2.2E-08	1.3E-08	9.3E-09	7.9E-09	6.4E-09
Gd-147	1.59 d	F	0.005	2.1E-09	5.0E-04	1.7E-09	8.4E-10	5.3E-10	3.1E-10	2.6E-10
		M	0.005	2.8E-09	5.0E-04	2.2E-09	1.1E-09	7.5E-10	5.1E-10	4.0E-10
Gd-148	93.0 y	F	0.005	8.3E-05	5.0E-04	7.6E-05	4.7E-05	3.2E-05	2.6E-05	2.6E-05
		M	0.005	3.2E-05	5.0E-04	2.9E-05	1.9E-05	1.3E-05	1.2E-05	1.1E-05
Gd-149	9.40 d	F	0.005	2.6E-09	5.0E-04	2.0E-09	8.0E-10	5.1E-10	3.1E-10	2.6E-10
		M	0.005	3.6E-09	5.0E-04	3.0E-09	1.5E-09	1.1E-09	9.2E-10	7.3E-10
Gd-151	120 d	F	0.005	6.3E-09	5.0E-04	4.9E-09	2.5E-09	1.5E-09	9.2E-10	7.8E-10
		M	0.005	4.5E-09	5.0E-04	3.5E-09	2.0E-09	1.3E-09	1.0E-09	8.6E-10
Gd-152	1.08E+14 y	F	0.005	5.9E-05	5.0E-04	5.4E-05	3.4E-05	2.4E-05	1.9E-05	1.9E-05
		M	0.005	2.1E-05	5.0E-04	1.9E-05	1.3E-05	8.9E-06	7.9E-06	8.0E-06
Gd-153	242 d	F	0.005	1.5E-08	5.0E-04	1.2E-08	6.5E-09	3.9E-09	2.4E-09	2.1E-09
		M	0.005	9.9E-09	5.0E-04	7.9E-09	4.8E-09	3.1E-09	2.5E-09	2.1E-09
Gd-159	18.6 h	F	0.005	1.2E-09	5.0E-04	8.9E-10	3.8E-10	2.3E-10	1.2E-10	1.0E-10
		M	0.005	2.2E-09	5.0E-04	1.5E-09	7.3E-10	4.9E-10	3.4E-10	2.7E-10
Terbium										
Tb-147	1.65 h	M	0.005	6.7E-10	5.0E-04	4.8E-10	2.3E-10	1.5E-10	9.3E-11	7.6E-11
Tb-149	4.15 h	M	0.005	2.1E-08	5.0E-04	1.5E-08	9.6E-09	6.6E-09	5.8E-09	4.9E-09
Tb-150	3.27 h	M	0.005	1.0E-09	5.0E-04	7.4E-10	3.5E-10	2.2E-10	1.3E-10	1.1E-10
Tb-151	17.6 h	M	0.005	1.6E-09	5.0E-04	1.2E-09	6.3E-10	4.2E-10	2.8E-10	2.3E-10
Tb-153	2.34 d	M	0.005	1.4E-09	5.0E-04	1.0E-09	5.4E-10	3.6E-10	2.3E-10	1.9E-10

Table A.2.—(continued)

Nuclide	Physical half-life	Type	f_1 <1y	$e(\tau)$ 3 Months	f_1 ≥1y	$e(\tau)$ 1 Year	5 Years	10 Years	15 Years	Adult
Tb-154	21.4 h	M	0.005	2.7E-09	5.0E-04	2.1E-09	1.1E-09	7.1E-10	4.5E-10	3.6E-10
Tb-155	5.32 d	M	0.005	1.4E-09	5.0E-04	1.0E-09	5.6E-10	3.4E-10	2.7E-10	2.2E-10
Tb-156	5.34 d	M	0.005	7.0E-09	5.0E-04	5.4E-09	3.0E-09	2.0E-09	1.5E-09	1.2E-09
Tb-156m	1.02 d	M	0.005	1.1E-09	5.0E-04	9.4E-10	4.7E-10	3.3E-10	2.7E-10	2.1E-10
Tb-156m	5.00 h	M	0.005	6.2E-10	5.0E-04	4.5E-10	2.4E-10	1.7E-10	1.2E-10	9.6E-11
Tb-157	1.50E+02 y	M	0.005	3.2E-09	5.0E-04	3.0E-09	2.0E-09	1.4E-09	1.2E-09	1.2E-09
Tb-158	1.50E+02 y	M	0.005	1.1E-07	5.0E-04	1.0E-07	7.0E-08	5.1E-08	4.7E-08	4.6E-08
Tb-160	72.3 d	M	0.005	3.2E-08	5.0E-04	2.5E-08	1.5E-08	1.0E-08	8.6E-09	7.0E-09
Tb-161	6.91 d	M	0.005	6.6E-09	5.0E-04	4.7E-09	2.6E-09	1.9E-09	1.6E-09	1.3E-09
Dysprosium										
Dy-155	10.0 h	M	0.005	5.6E-10	5.0E-04	4.4E-10	2.3E-10	1.5E-10	9.6E-11	7.7E-11
Dy-157	8.10 h	M	0.005	2.4E-10	5.0E-04	1.9E-10	9.9E-11	6.2E-11	3.8E-11	3.0E-11
Dy-159	144 d	M	0.005	2.1E-09	5.0E-04	1.7E-09	9.6E-10	6.0E-10	4.4E-10	3.7E-10
Dy-165	2.33 h	M	0.005	5.2E-10	5.0E-04	3.4E-10	1.6E-10	1.1E-10	7.2E-11	6.0E-11
Dy-166	3.40 d	M	0.005	1.2E-08	5.0E-04	8.3E-09	4.4E-09	3.0E-09	2.3E-09	1.9E-09
Holmium										
Ho-155	0.800 h	M	0.005	1.7E-10	5.0E-04	1.2E-10	5.8E-11	3.7E-11	2.4E-11	2.0E-11
Ho-157	0.210 h	M	0.005	3.4E-11	5.0E-04	2.5E-11	1.3E-11	8.0E-12	5.1E-12	4.2E-12
Ho-159	0.550 h	M	0.005	4.6E-11	5.0E-04	3.3E-11	1.7E-11	1.1E-11	7.5E-12	6.1E-12
Ho-161	2.50 h	M	0.005	5.7E-11	5.0E-04	4.0E-11	2.0E-11	1.2E-11	7.5E-12	6.0E-12
Ho-162	0.250 h	M	0.005	2.1E-11	5.0E-04	1.5E-11	7.2E-12	4.8E-12	3.4E-12	2.8E-12
Ho-162m	1.13 h	M	0.005	1.5E-10	5.0E-04	1.1E-10	5.8E-11	3.8E-11	2.6E-11	2.1E-11
Ho-164	0.483 h	M	0.005	6.8E-11	5.0E-04	4.5E-11	2.1E-11	1.4E-11	9.9E-12	8.4E-12
Ho-164m	0.625 h	M	0.005	9.1E-11	5.0E-04	5.9E-11	3.0E-11	2.0E-11	1.3E-11	1.2E-11
Ho-166	1.12 d	M	0.005	6.0E-09	5.0E-04	4.0E-09	1.9E-09	1.2E-09	7.9E-10	6.5E-10
Ho-166m	1.20E+03 y	M	0.005	2.6E-07	5.0E-04	2.5E-07	1.8E-07	1.3E-07	1.2E-07	1.2E-07
Ho-167	3.10 h	M	0.005	5.2E-10	5.0E-04	3.6E-10	1.8E-10	1.2E-10	8.7E-11	7.15-11
Erbium										
Er-161	3.24 h	M	0.005	3.8E-10	5.0E-04	2.9E-10	1.5E-10	9.5E-11	6.0E-11	4.8E-11
Er-165	10.4 h	M	0.005	7.2E-11	5.0E-04	5.3E-11	2.6E-11	1.6E-11	9.6E-12	7.9E-12
Er-169	9.30 d	M	0.005	4.7E-09	5.0E-04	3.5E-09	2.0E-09	1.5E-09	1.3E-09	1.0E-09
Er-171	7.52 h	M	0.005	1.8E-09	5.0E-04	1.2E-09	5.9E-10	3.9E-10	2.7E-10	2.2E-10
Er-172	2.05 d	M	0.005	6.6E-09	5.0E-04	4.7E-09	2.5E-09	1.7E-09	1.4E-09	1.1E-09

Table A.2.—*(continued)*

Nuclide	Physical half-life	Type	f_1 <1y	$e(\tau)$ 3 Months	f_1 ≥1y	$e(\tau)$ 1 Year	5 Years	10 Years	15 Years	Adult
Thulium										
Tm-162	0.362 h	M	0.005	1.3E-10	5.0E-04	9.6E-11	4.7E-11	3.0E-11	1.9E-11	1.6E-11
Tm-166	7.70 h	M	0.005	1.3E-09	5.0E-04	9.9E-10	5.2E-10	3.3E-10	2.2E-10	1.7E-10
Tm-167	9.24 d	M	0.005	5.6E-09	5.0E-04	4.1E-09	2.3E-09	1.7E-09	1.4E-09	1.1E-09
Tm-170	129 d	M	0.005	3.6E-08	5.0E-04	2.8E-08	1.6E-08	1.1E-08	8.5E-09	7.0E-09
Tm-171	1.92 y	M	0.005	6.8E-09	5.0E-04	5.7E-09	3.4E-09	2.0E-09	1.6E-09	1.4E-09
Tm-172	2.65 d	M	0.005	8.4E-09	5.0E-04	5.8E-09	2.9E-09	1.9E-09	1.4E-09	1.1E-09
Tm-173	8.24 h	M	0.005	1.5E-09	5.0E-04	1.0E-09	5.0E-10	3.3E-10	2.2E-10	1.8E-10
Tm-175	0.253 h	M	0.005	1.6E-10	5.0E-04	1.1E-10	5.0E-11	3.3E-11	2.2E-11	1.8E-11
Ytterbium										
Yb-162	0.315 h	M	0.005	1.1E-10	5.0E-04	7.9E-11	3.9E-11	2.5E-11	1.6E-11	1.3E-11
		S	0.005	1.2E-10	5.0E-04	8.2E-11	4.0E-11	2.6E-11	1.7E-11	1.4E-11
Yb-166	2.36 d	M	0.005	4.7E-09	5.0E-04	3.5E-09	1.9E-09	1.3E-09	9.0E-10	7.2E-10
		S	0.005	4.9E-09	5.0E-04	3.7E-09	2.0E-09	1.3E-09	9.6E-10	7.7E-10
Yb-167	0.292 h	M	0.005	4.4E-11	5.0E-04	3.1E-11	1.6E-11	1.1E-11	7.9E-12	6.5E-12
		S	0.005	4.6E-11	5.0E-04	3.2E-11	1.7E-11	1.1E-11	8.4E-12	6.9E-12
Yb-169	32.0 d	M	0.005	1.2E-08	5.0E-04	8.7E-09	5.1E-09	3.7E-09	3.2E-09	2.5E-09
		S	0.005	1.3E-08	5.0E-04	9.8E-09	5.9E-09	4.2E-09	3.7E-09	3.0E-09
Yb-175	4.19 d	M	0.005	3.5E-09	5.0E-04	2.5E-09	1.4E-09	9.8E-10	8.3E-10	6.5E-10
		S	0.005	3.7E-09	5.0E-04	2.7E-09	1.5E-09	1.1E-09	9.2E-10	7.3E-10
Yb-177	1.90 h	M	0.005	5.0E-10	5.0E-04	3.3E-10	1.6E-10	1.1E-10	7.8E-11	6.4E-11
		S	0.005	5.3E-10	5.0E-04	3.5E-10	1.7E-10	1.2E-10	8.4E-11	6.9E-11
Yb-178	1.23 h	M	0.005	5.9E-10	5.0E-04	3.9E-10	1.8E-10	1.2E-10	8.5E-11	7.0E-11
		S	0.005	6.2E-10	5.0E-04	4.1E-10	1.9E-10	1.3E-10	9.1E-11	7.5E-11
Lutetium										
Lu-169	1.42 d	M	0.005	2.3E-09	5.0E-04	1.8E-09	9.5E-10	6.3E-10	4.4E-10	3.5E-10
		S	0.005	2.4E-09	5.0E-04	1.9E-09	1.0E-09	6.7E-10	4.8E-10	3.8E-10
Lu-170	2.00 d	M	0.005	4.3E-09	5.0E-04	3.4E-09	1.8E-09	1.2E-09	7.8E-10	6.3E-10
		S	0.005	4.5E-09	5.0E-04	3.5E-09	1.8E-09	1.2E-09	8.2E-10	6.6E-10
Lu-171	8.22 d	M	0.005	5.0E-09	5.0E-04	3.7E-09	2.1E-09	1.2E-09	9.8E-10	8.0E-10
		S	0.005	4.7E-09	5.0E-04	3.9E-09	2.0E-09	1.4E-09	1.1E-09	8.8E-10
Lu-172	6.70 d	M	0.005	8.7E-09	5.0E-04	6.7E-09	3.8E-09	2.6E-09	1.8E-09	1.4E-09
		S	0.005	9.3E-09	5.0E-04	7.1E-09	4.0E-09	2.8E-09	2.0E-09	1.6E-09
Lu-173	1.37 y	M	0.005	1.0E-08	5.0E-04	8.5E-09	5.1E-09	3.2E-09	2.5E-09	2.2E-09
		S	0.005	1.0E-08	5.0E-04	8.7E-09	5.4E-09	3.6E-09	2.9E-09	2.4E-09
Lu-174	3.31 y	M	0.005	1.7E-08	5.0E-04	1.5E-08	9.1E-09	5.8E-09	4.7E-09	4.2E-09
		S	0.005	1.6E-08	5.0E-04	1.4E-08	8.9E-09	5.9E-09	4.9E-09	4.2E-09
Lu-174m	142 d	M	0.005	1.9E-08	5.0E-04	1.4E-08	8.6E-09	5.4E-09	4.3E-09	3.7E-09
		S	0.005	2.0E-08	5.0E-04	1.5E-08	9.2E-09	6.1E-09	5.0E-09	4.2E-09
Lu-176	3.60E+10 y	M	0.005	1.8E-07	5.0E-04	1.7E-07	1.1E-07	7.8E-08	7.1E-08	7.0E-08
		S	0.005	1.5E-07	5.0E-04	1.4E-07	9.4E-08	6.5E-08	5.9E-08	5.6E-08

Table A.2.—(*continued*)

Nuclide	Physical half-life	Type	f_1 <1y	$e(\tau)$ 3 Months	f_1 ≥1y	$e(\tau)$ 1 Year	5 Years	10 Years	15 Years	Adult
Lu-176m	3.68 h	M	0.005	8.9E-10	5.0E-04	5.9E-10	2.8E-10	1.9E-10	1.2E-10	1.1E-10
		S	0.005	9.3E-10	5.0E-04	6.2E-10	3.0E-10	2.0E-10	1.2E-10	1.2E-10
Lu-177	6.71 d	M	0.005	5.3E-09	5.0E-04	3.8E-09	2.2E-09	1.6E-09	1.4E-09	1.1E-09
		S	0.005	5.7E-09	5.0E-04	4.1E-09	2.4E-09	1.7E-09	1.5E-09	1.2E-09
Lu-177m	161 d	M	0.005	5.8E-08	5.0E-04	4.6E-08	2.8E-08	1.9E-08	1.6E-08	1.3E-08
		S	0.005	6.5E-08	5.0E-04	5.3E-08	3.2E-08	2.3E-08	2.0E-08	1.6E-08
Lu-178	0.473 h	M	0.005	2.3E-10	5.0E-04	1.5E-10	6.6E-11	4.3E-11	2.9E-11	2.4E-11
		S	0.005	2.4E-10	5.0E-04	1.5E-10	6.9E-11	4.5E-11	3.0E-11	2.6E-11
Lu-178m	0.378 h	M	0.005	2.6E-10	5.0E-04	1.8E-10	8.3E-11	5.6E-11	3.8E-11	3.2E-11
		S	0.005	2.7E-10	5.0E-04	1.9E-10	8.7E-11	5.8E-11	4.0E-11	3.3E-11
Lu-179	4.59 h	M	0.005	9.9E-10	5.0E-04	6.5E-10	3.0E-10	2.0E-10	1.2E-10	1.1E-10
		S	0.005	1.0E-09	5.0E-04	6.8E-10	3.2E-10	2.1E-10	1.3E-10	1.2E-10
Hafnium										
Hf-170	16.0 h	F	0.020	1.4E-09	0.002	1.1E-09	5.4E-10	3.4E-10	2.0E-10	1.6E-10
		M	0.020	2.2E-09	0.002	1.7E-09	8.7E-10	5.8E-10	3.9E-10	3.2E-10
Hf-172	1.87 y	F	0.020	1.5E-07	0.002	1.3E-07	7.8E-08	4.9E-08	3.5E-08	3.2E-08
		M	0.020	8.1E-08	0.002	6.9E-08	4.3E-08	2.8E-08	2.3E-08	2.0E-08
Hf-173	24.0 h	F	0.020	6.6E-10	0.002	5.0E-10	2.5E-10	1.5E-10	8.9E-11	7.4E-11
		M	0.020	1.1E-09	0.002	8.2E-10	4.3E-10	2.9E-10	2.0E-10	1.6E-10
Hf-175	70.0 d	F	0.020	5.4E-09	0.002	4.0E-09	2.1E-09	1.3E-09	8.5E-10	7.2E-10
		M	0.020	5.8E-09	0.002	4.5E-09	2.6E-09	1.8E-09	1.4E-09	1.2E-09
Hf-177m	0.856 h	F	0.020	3.9E-10	0.002	2.8E-10	1.3E-10	8.5E-11	5.2E-11	4.4E-11
		M	0.020	6.5E-10	0.002	4.7E-10	2.3E-10	1.5E-10	1.1E-10	9.0E-11
Hf-178m	31.0 y	F	0.020	6.2E-07	0.002	5.8E-07	4.0E-07	3.1E-07	2.7E-07	2.6E-07
		M	0.020	2.6E-07	0.002	2.4E-07	1.7E-07	1.3E-07	1.2E-07	1.2E-07
Hf-179m	25.1 d	F	0.020	9.7E-09	0.002	6.8E-09	3.4E-09	2.1E-09	1.2E-09	1.1E-09
		M	0.020	1.7E-08	0.002	1.3E-08	7.6E-09	5.5E-09	4.8E-09	3.8E-09
Hf-180m	5.50 h	F	0.020	5.4E-10	0.002	4.1E-10	2.0E-10	1.3E-10	7.2E-11	5.9E-11
		M	0.020	9.1E-10	0.002	6.8E-10	3.6E-10	2.4E-10	1.7E-10	1.3E-10
Hf-181	42.4 d	F	0.020	1.3E-08	0.002	9.6E-09	4.8E-09	2.8E-09	1.7E-09	1.4E-09
		M	0.020	2.2E-08	0.002	1.7E-08	9.9E-09	7.1E-09	6.3E-09	5.0E-09
Hf-182	9.00E+06 y	F	0.020	6.5E-07	0.002	6.2E-07	4.4E-07	3.6E-07	3.1E-07	3.1E-07
		M	0.020	2.4E-07	0.002	2.3E-07	1.7E-07	1.3E-07	1.3E-07	1.3E-07
Hf-182m	1.02 h	F	0.020	1.9E-10	0.002	1.4E-10	6.6E-11	4.2E-11	2.6E-11	2.1E-11
		M	0.020	3.2E-10	0.002	2.3E-10	1.2E-10	7.8E-11	5.6E-11	4.6E-11
Hf-183	1.07 h	F	0.020	2.5E-10	0.002	1.7E-10	7.9E-11	4.9E-11	2.8E-11	2.4E-11
		M	0.020	4.4E-10	0.002	3.0E-10	1.5E-10	9.8E-11	7.0E-11	5.7E-11
Hf-184	4.12 h	F	0.020	1.4E-09	0.002	9.6E-10	4.3E-10	2.7E-10	1.4E-10	1.2E-10
		M	0.020	2.6E-09	0.002	1.8E-09	8.9E-10	5.9E-10	4.0E-10	3.3E-10
Tantalum										
Ta-172	0.613 h	M	0.010	2.8E-10	0.001	1.9E-10	9.3E-11	6.0E-11	4.0E-11	3.3E-11
		S	0.010	2.9E-10	0.001	2.0E-10	9.8E-11	6.3E-11	4.2E-11	3.5E-11
Ta-173	3.65 h	M	0.010	8.8E-10	0.001	6.2E-10	3.0E-10	2.0E-10	1.3E-10	1.1E-10
		S	0.010	9.2E-10	0.001	6.5E-10	3.2E-10	2.1E-10	1.4E-10	1.1E-10

Table A.2.—(continued)

Nuclide	Physical half-life	Type	f_1 <1y	$e(\tau)$ 3 Months	f_1 ≥1y	$e(\tau)$ 1 Year	5 Years	10 Years	15 Years	Adult
Ta-174	1.20 h	M	0.010	3.2E-10	0.001	2.2E-10	1.1E-10	7.1E-11	5.0E-11	4.1E-11
		S	0.010	3.4E-10	0.001	2.3E-10	1.1E-10	7.5E-11	5.3E-11	4.3E-11
Ta-175	10.5 h	M	0.010	9.1E-10	0.001	7.0E-10	3.7E-10	2.4E-10	1.5E-10	1.2E-10
		S	0.010	9.5E-10	0.001	7.3E-10	3.8E-10	2.5E-10	1.6E-10	1.3E-10
Ta-176	8.08 h	M	0.010	1.4E-09	0.001	1.1E-09	5.7E-10	3.7E-10	2.4E-10	1.9E-10
		S	0.010	1.4E-09	0.001	1.1E-09	5.9E-10	3.8E-10	2.5E-10	2.0E-10
Ta-177	2.36 d	M	0.010	6.5E-10	0.001	4.7E-10	2.5E-10	1.5E-10	1.2E-10	9.6E-11
		S	0.010	6.9E-10	0.001	5.0E-10	2.7E-10	1.7E-10	1.3E-10	1.1E-10
Ta-178	2.20 h	M	0.010	4.4E-10	0.001	3.3E-10	1.7E-10	1.1E-10	8.0E-11	6.5E-11
		S	0.010	4.6E-10	0.001	3.4E-10	1.8E-10	1.2E-10	8.5E-11	6.8E-11
Ta-179	1.82 y	M	0.010	1.2E-09	0.001	9.6E-10	5.5E-10	3.5E-10	2.6E-10	2.2E-10
		S	0.010	2.4E-09	0.001	2.1E-09	1.3E-09	8.3E-10	6.4E-10	5.6E-10
Ta-180	1.00E+13 y	M	0.010	2.7E-08	0.001	2.2E-08	1.3E-08	9.2E-09	7.9E-09	6.4E-09
		S	0.010	7.0E-08	0.001	6.5E-08	4.5E-08	3.1E-08	2.8E-08	2.6E-08
Ta-180m	8.10 h	M	0.010	3.1E-10	0.001	2.2E-10	1.1E-10	7.4E-11	4.8E-11	4.4E-11
		S	0.010	3.3E-10	0.001	2.3E-10	1.2E-10	7.9E-11	5.2E-11	4.2E-11
Ta-182	115 d	M	0.010	3.2E-08	0.001	2.6E-08	1.5E-08	1.1E-08	9.5E-09	7.6E-09
		S	0.010	4.2E-08	0.001	3.4E-08	2.1E-08	1.5E-08	1.3E-08	1.0E-08
Ta-182m	0.264 h	M	0.010	1.6E-10	0.001	1.1E-10	4.9E-11	3.4E-11	2.4E-11	2.0E-11
		S	0.010	1.6E-10	0.001	1.1E-10	5.2E-11	3.6E-11	2.5E-11	2.1E-11
Ta-183	5.10 d	M	0.010	1.0E-08	0.001	7.4E-09	4.1E-09	2.9E-09	2.4E-09	1.9E-09
		S	0.010	1.1E-08	0.001	8.0E-09	4.5E-09	3.2E-09	2.7E-09	2.1E-09
Ta-184	8.70 h	M	0.010	3.2E-09	0.001	2.3E-09	1.1E-09	7.5E-10	5.0E-10	4.1E-10
		S	0.010	3.4E-09	0.001	2.4E-09	1.2E-09	7.9E-10	5.4E-10	4.3E-10
Ta-185	0.816 h	M	0.010	3.8E-10	0.001	2.5E-10	1.2E-10	7.7E-11	5.4E-11	4.5E-11
		S	0.010	4.0E-10	0.001	2.6E-10	1.2E-10	8.2E-11	5.7E-11	4.8E-11
Ta-186	0.175 h	M	0.010	1.6E-10	0.001	1.1E-10	4.8E-11	3.1E-11	2.0E-11	1.7E-11
		S	0.010	1.6E-10	0.001	1.1E-10	5.0E-11	3.2E-11	2.1E-11	1.8E-11
Tungsten										
W-176	2.30 h	F	0.600	3.3E-10	0.300	2.7E-10	1.4E-10	8.6E-11	5.0E-11	4.1E-11
W-177	2.25 h	F	0.600	2.0E-10	0.300	1.6E-10	8.2E-11	5.1E-11	3.0E-11	2.4E-11
W-178	21.7 d	F	0.600	7.2E-10	0.300	5.4E-10	2.5E-10	1.6E-10	8.7E-11	7.2E-11
W-179	0.625 h	F	0.600	9.3E-12	0.300	6.8E-12	3.3E-12	2.0E-12	1.2E-12	9.2E-13
W-181	121 d	F	0.600	2.5E-10	0.300	1.9E-10	9.2E-11	5.7E-11	3.2E-11	2.7E-11
W-185	75.1 d	F	0.600	1.4E-09	0.300	1.0E-09	4.4E-10	2.7E-10	1.4E-10	1.2E-10
W-187	23.9 h	F	0.600	2.0E-09	0.300	1.5E-09	7.0E-10	4.3E-10	2.3E-10	1.9E-10
W-188	69.4 d	F	0.600	7.1E-09	0.300	5.0E-09	2.2E-09	1.3E-09	6.8E-10	5.7E-10
Rhenium										
Re-177	0.233 h	F	1.000	9.4E-11	0.800	6.7E-11	3.2E-11	1.9E-11	1.2E-11	9.7E-12
		M	1.000	1.1E-10	0.800	7.9E-11	3.9E-11	2.5E-11	1.7E-11	1.4E-11
Re-178	0.220 h	F	1.000	9.9E-11	0.800	6.8E-11	3.1E-11	1.9E-11	1.2E-11	1.0E-11
		M	1.000	1.3E-10	0.800	8.5E-11	3.9E-11	2.6E-11	1.7E-11	1.4E-11

Table A.2.—(continued)

Nuclide	Physical half-life	Type	f_1 <1y	e(τ) 3 Months	f_1 ≥1y	e(τ) 1 Year	5 Years	10 Years	15 Years	Adult
Re-181	20.0 h	F	1.000	2.0E-09	0.800	1.4E-09	6.7E-10	3.8E-10	2.3E-10	1.8E-10
		M	1.000	2.1E-09	0.800	1.5E-09	7.4E-10	4.6E-10	3.1E-10	2.5E-10
Re-182	2.67 d	F	1.000	6.5E-09	0.800	4.7E-09	2.2E-09	1:3E-09	8.0E-10	6.4E-10
		M	1.000	8.7E-09	0.800	6.3E-09	3.4E-09	2.2E-09	1.5E-09	1.2E-09
Re-182	12.7 h	F	1.000	1.3E-09	0.800	1.0E-09	4.9E-10	2.8E-10	1.7E-10	1.4E-10
		M	1.000	1.4E-09	0.800	1.1E-09	5.7E-10	3.6E-10	2.5E-10	2.0E-10
Re-184	38.0 d	F	1.000	4.1E-09	0.800	2.9E-09	1.4E-09	8.6E-10	5.4E-10	4.4E-10
		M	1.000	9.1E-09	0.800	6.8E-09	4.0E-09	2.8E-09	2.4E-09	1.9E-09
Re-184m	165 d	F	1.000	6.6E-09	0.800	4.6E-09	2.0E-09	1.2E-09	7.3E-10	5.9E-10
		M	1.000	2.9E-08	0.800	2.2E-08	1.3E-08	9.3E-09	8.1E-09	6.5E-09
Re-186	3.78 d	F	1.000	7.3E-09	0.800	4.7E-09	2.0E-09	1.1E-09	6.6E-10	5.2E-10
		M	1.000	8.7E-09	0.800	5.7E-09	2.8E-09	1.8E-09	1.4E-09	1.1E-09
Re-186m	2.00E+05 y	F	1.000	1.2E-08	0.800	7.0E-09	2.9E-09	1.7E-09	1.0E-09	8.3E-10
		M	1.000	5.9E-08	0.800	4.6E-08	2.7E-08	1.8E-08	1.4E-08	1.2E-08
Re-187	5.00E+10 y	F	1.000	2.6E-11	0.800	1.6E-11	6.8E-12	3.8E-12	2.3E-12	1.8E-12
		M	1.000	5.7E-11	0.800	4.1E-11	2.0E-11	1.2E-11	7.5E-12	6.3E-12
Re-188	17.0 h	F	1.000	6.5E-09	0.800	4.4E-09	1.9E-09	1.0E-09	6.1E-10	4.6E-10
		M	1.000	6.0E-09	0.800	4.0E-09	1.8E-09	1.0E-09	6.8E-10	5.4E-10
Re-188m	0.310 h	F	1.000	1.4E-10	0.800	9.1E-11	4.0E-11	2.1E-11	1.3E-11	1.0E-11
		M	1.000	1.3E-10	0.800	8.6E-11	4.0E-11	2.7E-11	1.6E-11	1.3E-11
Re-189	1.01 d	F	1.000	3.7E-09	0.800	2.5E-09	1.1E-09	5.8E-10	3.5E-10	2.7E-10
		M	1.000	3.9E-09	0.800	2.6E-09	1.2E-09	7.6E-10	5.5E-10	4.3E-10
Osmium										
Os-180	0.366 h	F	0.020	7.1E-11	0.010	5.3E-11	2.6E-11	1.6E-11	1.0E-11	8.2E-12
		M	0.020	1.1E-10	0.010	7.9E-11	3.9E-11	2.5E-11	1.7E-11	1.4E-11
		S	0.020	1.1E-10	0.010	8.2E-11	4.1E-11	2.6E-11	1.8E-11	1.5E-11
Os-181	1.75 h	F	0.020	3.0E-10	0.010	2.3E-10	1.1E-10	7.0E-11	4.1E-11	3.3E-11
		M	0.020	4.5E-10	0.010	3.4E-10	1.8E-10	1.1E-10	7.6E-11	6.2E-11
		S	0.020	4.7E-10	0.010	3.6E-10	1.8E-10	1.2E-10	8.1E-11	6.5E-11
Os-182	22.0 h	F	0.020	1.6E-09	0.010	1.2E-09	6.0E-10	3.7E-10	2.1E-10	1.7E-10
		M	0.020	2.5E-09	0.010	1.9E-09	1.0E-09	6.6E-10	4.5E-10	3.6E-10
		S	0.020	2.6E-09	0.010	2.0E-09	1.0E-09	6.9E-10	4.8E-10	3.8E-10
Os-185	94.0 d	F	0.020	7.2E-09	0.010	5.8E-09	3.1E-09	1.9E-09	1.2E-09	1.1E-09
		M	0.020	6.6E-09	0.010	5.4E-09	2.9E-09	2.0E-09	1.5E-09	1.3E-09
		S	0.020	7.0E-09	0.010	5.8E-09	3.6E-09	2.4E-09	1.9E-09	1.6E-09
Os-189m	6.00 h	F	0.020	3.8E-11	0.010	2.8E-11	1.2E-11	7.0E-12	3.5E-12	2.5E-12
		M	0.020	6.5E-11	0.010	4.1E-11	1.8E-11	1.1E-11	6.0E-12	5.0E-12
		S	0.020	6.8E-11	0.010	4.3E-11	1.9E-11	1.2E-11	6.3E-12	5.3E-12
Os-191	15.4 d	F	0.020	2.8E-09	0.010	1.9E-09	8.5E-10	5.3E-10	3.0E-10	2.5E-10
		M	0.020	8.0E-09	0.010	5.8E-09	3.4E-09	2.4E-09	2.0E-09	1.7E-09
		S	0.020	9.0E-09	0.010	6.5E-09	3.9E-09	2.7E-09	2.3E-09	1.9E-09
Os-191m	13.0 h	F	0.020	3.0E-10	0.010	2.0E-10	8.8E-11	5.4E-11	2.9E-11	2.4E-11
		M	0.020	7.8E-10	0.010	5.4E-10	3.1E-10	2.1E-10	1.7E-10	1.4E-10
		S	0.020	8.5E-10	0.010	6.0E-10	3.4E-10	2.4E-10	2.0E-10	1.6E-10
Os-193	1.25 d	F	0.020	1.9E-09	0.010	1.2E-09	5.2E-10	3.2E-10	1.8E-10	1.6E-10
		M	0.020	3.8E-09	0.010	2.6E-09	1.3E-09	8.4E-10	5.9E-10	4.8E-10
		S	0.020	4.0E-09	0.010	2.7E-09	1.3E-09	9.0E-10	6.4E-10	5.2E-10

Table A.2.—(continued)

Nuclide	Physical half-life	Type	f_1 <1y	$e(\tau)$ 3 Months	f_1 ≥1y	$e(\tau)$ 1 Year	5 Years	10 Years	15 Years	Adult
Os-194	6.00 y	F	0.020	8.7E-08	0.010	6.8E-08	3.4E-08	2.1E-08	1.3E-08	1.1E-08
		M	0.020	9.9E-08	0.010	8.3E-08	4.8E-08	3.1E-08	2.4E-08	2.1E-08
		S	0.020	2.6E-07	0.010	2.4E-07	1.6E-07	1.1E-07	8.8E-08	8.5E-08
Iridium										
Ir-182	0.250 h	F	0.020	1.4E-10	0.010	9.8E-11	4.5E-11	2.8E-11	1.7E-11	1.4E-11
		M	0.020	2.1E-10	0.010	1.4E-10	6.7E-11	4.3E-11	2.8E-11	2.3E-11
		S	0.020	2.2E-10	0.010	1.5E-10	6.9E-11	4.4E-11	2.9E-11	2.4E-11
Ir-184	3.02 h	F	0.020	5.7E-10	0.010	4.4E-10	2.1E-10	1.3E-10	7.6E-11	6.2E-11
		M	0.020	8.6E-10	0.010	6.4E-10	3.2E-10	2.1E-10	1.4E-10	1.1E-10
		S	0.020	8.9E-10	0.010	6.6E-10	3.4E-10	2.2E-10	1.4E-10	1.2E-10
Ir-185	14.0 h	F	0.020	8.0E-10	0.010	6.1E-10	2.9E-10	1.8E-10	1.0E-10	8.2E-11
		M	0.020	1.3E-09	0.010	9.7E-10	4.9E-10	3.2E-10	2.2E-10	1.8E-10
		S	0.020	1.4E-09	0.010	1.0E-09	5.2E-10	3.4E-10	2.3E-10	1.9E-10
Ir-186	15.8 h	F	0.020	1.5E-09	0.010	1.2E-09	5.9E-10	3.6E-10	2.1E-10	1.7E-10
		M	0.020	2.2E-09	0.010	1.7E-09	8.8E-10	5.8E-10	3.8E-10	3.1E-10
		S	0.020	2.3E-09	0.010	1.8E-09	9.2E-10	6.0E-10	4.0E-10	3.2E-10
Ir-186	1.75 h	F	0.020	2.1E-10	0.010	1.6E-10	7.7E-11	4.8E-11	2.8E-11	2.3E-11
		M	0.020	3.3E-10	0.010	2.4E-10	1.2E-10	7.7E-11	5.1E-11	4.2E-11
		S	0.020	3.4E-10	0.010	2.5E-10	1.2E-10	8.1E-11	5.4E-11	4.4E-11
Ir-187	10.5 h	F	0.020	3.6E-10	0.010	2.8E-10	1.4E-10	8.2E-11	4.6E-11	3.7E-11
		M	0.020	5.8E-10	0.010	4.3E-10	2.2E-10	1.4E-10	9.2E-11	7.4E-11
		S	0.020	6.0E-10	0.010	4.5E-10	2.3E-10	1.5E-10	9.7E-11	7.9E-11
Ir-188	1.73 d	F	0.020	2.0E-09	0.010	1.6E-09	8.0E-10	5.0E-10	2.9E-10	2.4E-10
		M	0.020	2.7E-09	0.010	2.1E-09	1.1E-09	7.5E-10	5.0E-10	4.0E-10
		S	0.020	2.8E-09	0.010	2.2E-09	1.2E-09	7.8E-10	5.2E-10	4.2E-10
Ir-189	13.3 d	F	0.020	1.2E-09	0.010	8.2E-10	3.8E-10	2.4E-10	1.3E-10	1.1E-10
		M	0.020	2.7E-09	0.010	1.9E-09	1.1E-09	7.7E-10	6.4E-10	5.2E-10
		S	0.020	3.0E-09	0.010	2.2E-09	1.3E-09	8.7E-10	7.3E-10	6.0E-10
Ir-190	12.1 d	F	0.020	6.2E-09	0.010	4.7E-09	2.4E-09	1.5E-09	9.1E-10	7.7E-10
		M	0.020	1.1E-08	0.010	8.6E-09	4.4E-09	3.1E-09	2.7E-09	2.1E-09
		S	0.020	1.1E-08	0.010	9.4E-09	4.8E-09	3.5E-09	3.0E-09	2.4E-09
Ir-190m	3.10 h	F	0.020	4.2E-10	0.010	3.4E-10	1.7E-10	1.0E-10	6.0E-11	4.9E-11
		M	0.020	6.0E-10	0.010	4.7E-10	2.4E-10	1.5E-10	9.9E-11	7.9E-11
		S	0.020	6.2E-10	0.010	4.8E-10	2.5E-10	1.6E-10	1.0E-10	8.3E-11
Ir-190m	1.20 h	F	0.020	3.2E-11	0.010	2.4E-11	1.2E-11	7.2E-12	4.3E-12	3.6E-12
		M	0.020	5.7E-11	0.010	4.2E-11	2.0E-11	1.4E-11	1.2E-11	9.3E-12
		S	0.020	5.5E-11	0.010	4.5E-11	2.2E-11	1.6E-11	1.3E-11	1.0E-11
Ir-192	74.0 d	F	0.020	1.5E-08	0.010	1.1E-08	5.7E-09	3.3E-09	2.1E-09	1.8E-09
		M	0.020	2.3E-08	0.010	1.8E-08	1.1E-08	7.6E-09	6.4E-09	5.2E-09
		S	0.020	2.8E-08	0.010	2.2E-08	1.3E-08	9.5E-09	8.1E-09	6.6E-09
Ir-192m	2.41E+02 y	F	0.020	2.7E-08	0.010	2.3E-08	1.4E-08	8.2E-09	5.4E-09	4.8E-09
		M	0.020	2.3E-08	0.010	2.1E-08	1.3E-08	8.4E-09	6.6E-09	5.8E-09
		S	0.020	9.2E-08	0.010	9.1E-08	6.5E-08	4.5E-08	4.0E-08	3.9E-08
Ir-193m	11.9 d	F	0.020	1.2E-09	0.010	8.4E-10	3.7E-10	2.2E-10	1.2E-10	1.0E-10
		M	0.020	4.8E-09	0.010	3.5E-09	2.1E-09	1.5E-09	1.4E-09	1.1E-09
		S	0.020	5.4E-09	0.010	4.0E-09	2.4E-09	1.8E-09	1.6E-09	1.3E-09
Ir-194	19.1 h	F	0.020	2.9E-09	0.010	1.9E-09	8.1E-10	4.9E-10	2.5E-10	2.1E-10
		M	0.020	5.3E-09	0.010	3.5E-09	1.6E-09	1.0E-09	6.3E-10	5.2E-10
		S	0.020	5.5E-09	0.010	3.7E-09	1.7E-09	1.1E-09	6.7E-10	5.6E-10

Table A.2.—(continued)

Nuclide	Physical half-life	Type	f_1 <1y	e(τ) 3 Months	f_1 ≥1y	e(τ) 1 Year	5 Years	10 Years	15 Years	Adult
Ir-194m	171 d	F	0.020	3.4E-08	0.010	2.7E-08	1.4E-08	9.5E-09	6.2E-09	5.4E-09
		M	0.020	3.9E-08	0.010	3.2E-08	1.9E-08	1.3E-08	1.1E-08	9.0E-09
		S	0.020	5.0E-08	0.010	4.2E-08	2.6E-08	1.8E-08	1.5E-08	1.3E-08
Ir-195	2.50 h	F	0.020	2.9E-10	0.010	1.9E-10	8.1E-11	5.1E-11	2.9E-11	2.4E-11
		M	0.020	5.4E-10	0.010	3.6E-10	1.7E-10	1.1E-10	8.1E-11	6.7E-11
		S	0.020	5.7E-10	0.010	3.8E-10	1.8E-10	1.2E-10	8.7E-11	7.1E-11
Ir-195m	3.80 h	F	0.020	6.9E-10	0.010	4.8E-10	2.1E-10	1.3E-10	7.2E-11	6.0E-11
		M	0.020	1.2E-09	0.010	8.6E-10	4.2E-10	2.7E-10	1.9E-10	1.6E-10
		S	0.020	1.3E-09	0.010	9.0E-10	4.4E-10	2.9E-10	2.0E-10	1.7E-10
Platinum										
Pt-186	2.00 h	F	0.020	3.0E-10	0.010	2.4E-10	1.2E-10	7.2E-11	4.1E-11	3.3E-11
Pt-188	10.2 d	F	0.020	3.6E-09	0.010	2.7E-09	1.3E-09	8.4E-10	5.0E-10	4.2E-10
Pt-189	10.9 h	F	0.020	3.8E-10	0.010	2.9E-10	1.4E-10	8.4E-11	4.7E-11	3.8E-11
Pt-191	2.80 d	F	0.020	1.1E-09	0.010	7.9E-10	3.7E-10	2.3E-10	1.3E-10	1.1E-10
Pt-193	50.0 y	F	0.020	2.2E-10	0.010	1.6E-10	7.2E-11	4.3E-11	2.5E-11	2.1E-11
Pt-193m	4.33 d	F	0.020	1.6E-09	0.010	1.0E-09	4.5E-10	2.7E-10	1.4E-10	1.2E-10
Pt-195m	4.02 d	F	0.020	2.2E-09	0.010	1.5E-09	6.4E-10	3.9E-10	2.1E-10	1.8E-10
Pt-197	18.3 h	F	0.020	1.1E-09	0.010	7.3E-10	3.1E-10	1.9E-10	1.0E-10	8.5E-11
Pt-197m	1.57 h	F	0.020	2.8E-10	0.010	1.8E-10	7.9E-11	4.9E-11	2.8E-11	2.4E-11
Pt-199	0.513 h	F	0.020	1.3E-10	0.010	8.3E-11	3.6E-11	2.3E-11	1.4E-11	1.2E-11
Pt-200	12.5 h	F	0.020	2.6E-09	0.010	1.7E-09	7.2E-10	5.1E-10	2.6E-10	2.2E-10
Gold										
Au-193	17.6 h	F	0.200	3.7E-10	0.100	2.8E-10	1.3E-10	7.9E-11	4.3E-11	3.6E-11
		M	0.200	7.5E-10	0.100	5.6E-10	2.8E-10	1.9E-10	1.4E-10	1.1E-10
		S	0.200	7.9E-10	0.100	5.9E-10	3.0E-10	2.0E-10	1.5E-10	1.2E-10
Au-194	1.65 d	F	0.200	1.2E-09	0.100	9.6E-10	4.9E-10	3.0E-10	1.8E-10	1.4E-10
		M	0.200	1.7E-09	0.100	1.4E-09	7.1E-10	4.6E-10	2.9E-10	2.3E-10
		S	0.200	1.7E-09	0.100	1.4E-09	7.3E-10	4.7E-10	3.0E-10	2.4E-10
Au-195	183 d	F	0.200	7.2E-10	0.100	5.3E-10	2.5E-10	1.5E-10	8.1E-11	6.6E-11
		M	0.200	5.2E-09	0.100	4.1E-09	2.4E-09	1.6E-09	1.4E-09	1.1E-09
		S	0.200	8.1E-09	0.100	6.6E-09	3.9E-09	2.6E-09	2.1E-09	1.7E-09
Au-198	2.69 d	F	0.200	2.4E-09	0.100	1.7E-09	7.6E-10	4.7E-10	2.5E-10	2.1E-10
		M	0.200	5.0E-09	0.100	4.1E-09	1.9E-09	1.3E-09	9.7E-10	7.8E-10
		S	0.200	5.4E-09	0.100	4.4E-09	2.0E-09	1.4E-09	1.1E-09	8.6E-10
Au-198m	2.30 d	F	0.200	3.3E-09	0.100	2.4E-09	1.1E-09	6.9E-10	3.7E-10	3.2E-10
		M	0.200	8.7E-09	0.100	6.5E-09	3.6E-09	2.6E-09	2.2E-09	1.8E-09
		S	0.200	9.5E-09	0.100	7.1E-09	4.0E-09	2.9E-09	2.5E-09	2.0E-09
Au-199	3.14 d	F	0.200	1.1E-09	0.100	7.9E-10	3.5E-10	2.2E-10	1.1E-10	9.8E-11
		M	0.200	3.4E-09	0.100	2.5E-09	1.4E-09	1.0E-09	9.0E-10	7.1E-10
		S	0.200	3.8E-09	0.100	2.8E-09	1.6E-09	1.2E-09	1.0E-09	7.9E-10
Au-200	0.807 h	F	0.200	1.9E-10	0.100	1.2E-10	5.2E-11	3.2E-11	1.9E-11	1.6E-11
		M	0.200	3.2E-10	0.100	2.1E-10	9.3E-11	6.0E-11	4.0E-11	3.3E-11
		S	0.200	3.4E-10	0.100	2.1E-10	9.8E-11	6.3E-11	4.2E-11	3.5E-11

Table A.2.—*(continued)*

Nuclide	Physical half-life	Type	f_1 <1y	e(τ) 3 Months	f_1 ≥1y	e(τ) 1 Year	5 Years	10 Years	15 Years	Adult
Au-200m	18.7 h	F	0.200	2.7E-09	0.100	2.1E-09	1.0E-09	6.4E-10	3.6E-10	2.9E-10
		M	0.200	4.8E-09	0.100	3.7E-09	1.9E-09	1.2E-09	8.4E-10	6.8E-10
		S	0.200	5.1E-09	0.100	3.9E-09	2.0E-09	1.3E-09	8.9E-10	7.2E-10
Au-201	0.440 h	F	0.200	9.0E-11	0.100	5.7E-11	2.5E-11	1.6E-11	1.0E-11	8.7E-12
		M	0.200	1.5E-10	0.100	9.6E-11	4.3E-11	2.9E-11	2.0E-11	1.7E-11
		S	0.200	1.5E-10	0.100	1.0E-10	4.5E-11	3.0E-11	2.1E-11	1.7E-11
Mercury										
Hg-193 (organic)	3.50 h	F	0.800	2.2E-10	0.400	1.8E-10	8.2E-11	5.0E-11	2.9E-11	2.4E-11
Hg-193 (inorganic)	3.50 h	F	0.040	2.7E-10	0.020	2.0E-10	8.9E-11	5.5E-11	3.1E-11	2.6E-11
		M	0.040	5.3E-10	0.020	3.8E-10	1.9E-10	1.3E-10	9.2E-11	7.5E-11
Hg-193m (organic)	11.1 h	F	0.800	8.4E-10	0.400	7.6E-10	3.7E-10	2.2E-10	1.3E-10	1.0E-10
Hg-193m (inorganic)	11.1 h	F	0.040	1.1E-09	0.020	8.5E-10	4.1E-10	2.5E-10	1.4E-10	1.1E-10
		M	0.040	1.9E-09	0.020	1.4E-09	7.2E-10	4.7E-10	3.2E-10	2.6E-10
Hg-194 (organic)	2.60E+02 y	F	0.800	4.9E-08	0.400	3.7E-08	2.4E-08	1.9E-08	1.5E-08	1.4E-08
Hg-194 (inorganic)	2.60E+02 y	F	0.040	3.2E-08	0.020	2.9E-08	2.0E-08	1.6E-08	1.4E-08	1.3E-08
		M	0.040	2.1E-08	0.020	1.9E-08	1.3E-08	1.0E-08	8.9E-09	8.3E-09
Hg-195 (organic)	9.90 h	F	0.800	2.0E-10	0.400	1.8E-10	8.5E-11	5.1E-11	2.8E-11	2.3E-11
Hg-195 (inorganic)	9.90 h	F	0.040	2.7E-10	0.020	2.0E-10	9.5E-11	5.7E-11	3.1E-11	2.5E-11
		M	0.040	5.3E-10	0.020	3.9E-10	2.0E-10	1.3E-10	9.0E-11	7.3E-11
Hg-195m (organic)	1.73 d	F	0.800	1.1E-09	0.400	9.7E-10	4.4E-10	2.7E-10	1.4E-10	1.2E-10
Hg-195m (inorganic)	1.73 d	F	0.040	1.6E-09	0.020	1.1E-09	5.1E-10	3.1E-10	1.7E-10	1.4E-10
		M	0.040	3.7E-09	0.020	2.6E-09	1.4E-09	8.5E-10	6.7E-10	5.3E-10
Hg-197 (organic)	2.67 d	F	0.800	4.7E-10	0.400	4.0E-10	1.8E-10	1.1E-10	5.8E-11	4.7E-11
Hg-197 (inorganic)	2.67 d	F	0.040	6.8E-10	0.020	4.7E-10	2.1E-10	1.3E-10	6.8E-11	5.6E-11
		M	0.040	1.7E-09	0.020	1.2E-09	6.6E-10	4.6E-10	3.8E-10	3.0E-10
Hg-197m (organic)	23.8 h	F	0.800	9.3E-10	0.400	7.8E-10	3.4E-10	2.1E-10	1.1E-10	9.6E-11
Hg-197m (inorganic)	23.8 h	F	0.040	1.4E-09	0.020	9.3E-10	4.0E-10	2.5E-10	1.3E-10	1.1E-10
		M	0.040	3.5E-09	0.020	2.5E-09	1.1E-09	8.2E-10	6.7E-10	5.3E-10
Hg-199m (organic)	0.710 h	F	0.800	1.4E-10	0.400	9.6E-11	4.2E-11	2.7E-11	1.7E-11	1.5E-11
Hg-199m (inorganic)	0.710 h	F	0.040	1.4E-10	0.020	9.6E-11	4.2E-11	2.7E-11	1.7E-11	1.5E-11
		M	0.040	2.5E-10	0.020	1.7E-10	7.9E-11	5.4E-11	3.8E-11	3.2E-11
Hg-203 (organic)	46.6 d	F	0.800	5.7E-09	0.400	3.7E-09	1.7E-09	1.1E-09	6.6E-10	5.6E-10
Hg-203 (inorganic)	46.6 d	F	0.040	4.2E-09	0.020	2.9E-09	1.4E-09	9.0E-10	5.5E-10	4.6E-10
		M	0.040	1.0E-08	0.020	7.9E-09	4.7E-09	3.4E-09	3.0E-09	2.4E-09
Thallium										
Tl-194	0.550 h	F	1.000	3.6E-11	1.000	3.0E-11	1.5E-11	9.2E-12	5.5E-12	4.4E-12

Table A.2.—(continued)

Nuclide	Physical half-life	Type	f_1 <1y	$e(\tau)$ 3 Months	f_1 ≥1y	$e(\tau)$ 1 Year	5 Years	10 Years	15 Years	Adult
Tl-194m	0.546 h	F	1.000	1.7E-10	1.000	1.2E-10	6.1E-11	3.8E-11	2.3E-11	1.9E-11
Tl-195	1.16 h	F	1.000	1.3E-10	1.000	1.0E-10	5.3E-11	3.2E-11	1.9E-11	1.5E-11
Tl-197	2.84 h	F	1.000	1.3E-10	1.000	9.7E-11	4.7E-11	2.9E-11	1.7E-11	1.4E-11
Tl-198	5.30 h	F	1.000	4.7E-10	1.000	4.0E-10	2.1E-10	1.3E-10	7.5E-11	6.0E-11
Tl-198m	1.87 h	F	1.000	3.2E-10	1.000	2.5E-10	1.2E-10	7.5E-11	4.5E-11	3.7E-11
Tl-199	7.42 h	F	1.000	1.7E-10	1.000	1.3E-10	6.4E-11	3.9E-11	2.3E-11	1.9E-11
Tl-200	1.09 d	F	1.000	1.0E-09	1.000	8.7E-10	4.6E-10	2.8E-10	1.6E-10	1.3E-10
Tl-201	3.04 d	F	1.000	4.5E-10	1.000	3.3E-10	1.5E-10	9.4E-11	5.4E-11	4.4E-11
Tl-202	12.2 d	F	1.000	1.5E-09	1.000	1.2E-09	5.9E-10	3.8E-10	2.3E-10	1.9E-10
Tl-204	3.78 y	F	1.000	5.0E-09	1.000	3.3E-09	1.5E-09	8.8E-10	4.7E-10	3.9E-10
Lead[a,b]										
Pb-195m	0.263 h	F	0.600	1.3E-10	0.200	1.0E-10	4.9E-11	3.1E-11	1.9E-11	1.6E-11
		M	0.200	2.0E-10	0.100	1.5E-10	7.1E-11	4.6E-11	3.1E-11	2.5E-11
		S	0.020	2.1E-10	0.010	1.5E-10	7.4E-11	4.8E-11	3.2E-11	2.7E-11
Pb-198	2.40 h	F	0.600	3.4E-10	0.200	2.9E-10	1.5E-10	8.9E-11	5.2E-11	4.3E-11
		M	0.200	5.0E-10	0.100	4.0E-10	2.1E-10	1.3E-10	8.3E-11	6.6E-11
		S	0.020	5.4E-10	0.010	4.2E-10	2.2E-10	1.4E-10	8.7E-11	7.0E-11
Pb-199	1.50 h	F	0.600	1.9E-10	0.200	1.6E-10	8.2E-11	4.9E-11	2.9E-11	2.3E-11
		M	0.200	2.8E-10	0.100	2.2E-10	1.1E-10	7.1E-11	4.5E-11	3.6E-11
		S	0.020	2.9E-10	0.010	2.3E-10	1.2E-10	7.4E-11	4.7E-11	3.7E-11
Pb-200	21.5 h	F	0.600	1.1E-09	0.200	9.3E-10	4.6E-10	2.8E-10	1.6E-10	1.4E-10
		M	0.200	2.2E-09	0.100	1.7E-09	8.6E-10	5.7E-10	4.1E-10	3.3E-10
		S	0.020	2.4E-09	0.010	1.8E-09	9.2E-10	6.2E-10	4.4E-10	3.5E-10
Pb-201	9.40 h	F	0.600	4.8E-10	0.200	4.1E-10	2.0E-10	1.2E-10	7.1E-11	6.0E-11
		M	0.200	8.0E-10	0.100	6.4E-10	3.3E-10	2.1E-10	1.4E-10	1.1E-10
		S	0.020	8.8E-10	0.010	6.7E-10	3.5E-10	2.2E-10	1.5E-10	1.2E-10
Pb-202	3.00E+05 y	F	0.600	1.9E-08	0.200	1.3E-08	8.9E-09	1.3E-08	1.8E-08	1.1E-08
		M	0.200	1.2E-08	0.100	8.9E-09	6.2E-09	6.7E-09	8.7E-09	6.3E-09
		S	0.020	2.8E-08	0.010	2.8E-08	2.0E-08	1.4E-08	1.3E-08	1.2E-08
Pb-202m	3.62 h	F	0.600	4.7E-10	0.200	4.0E-10	2.1E-10	1.3E-10	7.5E-11	6.2E-11
		M	0.200	6.9E-10	0.100	5.6E-10	2.9E-10	1.9E-10	1.2E-10	9.5E-11
		S	0.020	7.3E-10	0.010	5.8E-10	3.0E-10	1.9E-10	1.3E-10	1.0E-10
Pb-203	2.17 d	F	0.600	7.2E-10	0.200	5.8E-10	2.8E-10	1.7E-10	9.9E-11	8.5E-11
		M	0.200	1.3E-09	0.100	1.0E-09	5.4E-10	3.6E-10	2.5E-10	2.0E-10
		S	0.020	1.5E-09	0.010	1.1E-09	5.8E-10	3.8E-10	2.8E-10	2.2E-10
Pb-205	1.43E+07 y	F	0.600	1.1E-09	0.200	6.9E-10	4.0E-10	4.1E-10	4.3E-10	3.3E-10
		M	0.200	1.1E-09	0.100	7.7E-10	4.3E-10	3.2E-10	2.9E-10	2.5E-10
		S	0.020	2.9E-09	0.010	2.7E-09	1.7E-09	1.1E-09	9.2E-10	8.5E-10
Pb-209	3.25 h	F	0.600	1.8E-10	0.200	1.2E-10	5.3E-11	3.4E-11	1.9E-11	1.7E-11
		M	0.200	4.0E-10	0.100	2.7E-10	1.3E-10	9.2E-11	6.9E-11	5.6E-11
		S	0.020	4.4E-10	0.010	2.9E-10	1.4E-10	9.9E-11	7.5E-11	6.1E-11
Pb-210	22.3 y	F	0.600	4.7E-06	0.200	2.9E-06	1.5E-06	1.4E-06	1.3E-06	9.0E-07
		M	0.200	5.0E-06	0.100	3.7E-06	2.2E-06	1.5E-06	1.3E-06	1.1E-06
		S	0.020	1.8E-05	0.010	1.8E-05	1.1E-05	7.2E-06	5.9E-06	5.6E-06

a Dose coefficients for this element are based on age-specific biokinetic data

b The f_1 value for 1 to 15 year olds for Type F is 0.4

Table A.2.—(*continued*)

Nuclide	Physical half-life	Type	f_1 <1y	$e(\tau)$ 3 Months	f_1 ≥1y	$e(\tau)$ 1 Year	5 Years	10 Years	15 Years	Adult
Pb-211	0.601 h	F	0.600	2.5E-08	0.200	1.7E-08	8.7E-09	6.1E-09	4.6E-09	3.9E-09
		M	0.200	6.2E-08	0.100	4.5E-08	2.5E-08	1.9E-08	1.4E-08	1.1E-08
		S	0.020	6.6E-08	0.010	4.8E-08	2.7E-08	2.0E-08	1.5E-08	1.2E-08
Pb-212	10.6 h	F	0.600	1.9E-07	0.200	1.2E-07	5.4E-08	3.5E-08	2.0E-08	1.8E-08
		M	0.200	6.2E-07	0.100	4.6E-07	3.0E-07	2.2E-07	2.2E-07	1.7E-07
		S	0.020	6.7E-07	0.010	5.0E-07	3.3E-07	2.5E-07	2.4E-07	1.9E-07
Pb-214	0.447 h	F	0.600	2.2E-08	0.200	1.5E-08	6.9E-09	4.8E-09	3.3E-09	2.8E-09
		M	0.200	6.4E-08	0.100	4.6E-08	2.6E-08	1.9E-08	1.4E-08	1.4E-08
		S	0.020	6.9E-08	0.010	5.0E-08	2.8E-08	2.1E-08	1.5E-08	1.5E-08
Bismuth										
Bi-200	0.606 h	F	0.100	1.9E-10	0.050	1.5E-10	7.4E-11	4.5E-11	2.7E-11	2.2E-11
		M	0.100	2.5E-10	0.050	1.9E-10	9.9E-11	6.3E-11	4.1E-11	3.3E-11
Bi-201	1.80 h	F	0.100	4.0E-10	0.050	3.1E-10	1.5E-10	9.3E-11	5.4E-11	4.4E-11
		M	0.100	5.5E-10	0.050	4.1E-10	2.0E-10	1.3E-10	8.3E-11	6.6E-11
Bi-202	1.67 h	F	0.100	3.4E-10	0.050	2.8E-10	1.5E-10	9.0E-11	5.3E-11	4.3E-11
		M	0.100	4.2E-10	0.050	3.4E-10	1.8E-10	1.1E-10	6.9E-11	5.5E-11
Bi-203	11.8 h	F	0.100	1.5E-09	0.050	1.2E-09	6.4E-10	4.0E-10	2.3E-10	1.9E-10
		M	0.100	2.0E-09	0.050	1.6E-09	8.2E-10	5.3E-10	3.3E-10	2.6E-10
Bi-205	15.3 d	F	0.100	3.0E-09	0.050	2.4E-09	1.3E-09	8.0E-10	4.7E-10	3.8E-10
		M	0.100	5.5E-09	0.050	4.4E-09	2.5E-09	1.6E-09	1.2E-09	9.3E-10
Bi-206	6.24 d	F	0.100	6.1E-09	0.050	4.8E-09	2.5E-09	1.6E-09	9.1E-10	7.4E-10
		M	0.100	1.0E-08	0.050	8.0E-09	4.4E-09	2.9E-09	2.1E-09	1.7E-09
Bi-207	38.0 y	F	0.100	4.3E-09	0.050	3.3E-09	1.7E-09	1.0E-09	6.0E-10	4.9E-10
		M	0.100	2.3E-08	0.050	2.0E-08	1.2E-08	8.2E-09	6.5E-09	5.6E-09
Bi-210	5.01 d	F	0.100	1.1E-08	0.050	6.9E-09	3.2E-09	2.1E-09	1.3E-09	1.1E-09
		M	0.100	3.9E-07	0.050	3.0E-07	1.9E-07	1.3E-07	1.1E-07	9.3E-08
Bi-210m	3.00E+06 y	F	0.100	4.1E-07	0.050	2.6E-07	1.3E-07	8.3E-08	5.6E-08	4.6E-08
		M	0.100	1.5E-05	0.050	1.1E-05	7.0E-06	4.8E-06	4.1E-06	3.4E-06
Bi-212	1.01 h	F	0.100	6.5E-08	0.050	4.5E-08	2.1E-08	1.5E-08	1.0E-08	9.1E-09
		M	0.100	1.6E-07	0.050	1.1E-07	6.0E-08	4.4E-08	3.8E-08	3.1E-08
Bi-213	0.761 h	F	0.100	7.7E-08	0.050	5.3E-08	2.5E-08	1.7E-08	1.2E-08	1.0E-08
		M	0.100	1.6E-07	0.050	1.2E-07	6.0E-08	4.4E-08	3.6E-08	3.0E-08
Bi-214	0.332 h	F	0.100	5.0E-08	0.050	3.5E-08	1.6E-08	1.1E-08	8.2E-09	7.1E-09
		M	0.100	8.7E-08	0.050	6.1E-08	3.1E-08	2.2E-08	1.7E-08	1.4E-08
Polonium[a]										
Po-203	0.612 h	F	0.200	1.9E-10	0.100	1.5E-10	7.7E-11	4.7E-11	2.8E-11	2.3E-11
		M	0.200	2.7E-10	0.100	2.1E-10	1.1E-10	6.7E-11	4.3E-11	3.5E-11
		S	0.020	2.8E-10	0.010	2.2E-10	1.1E-10	7.0E-11	4.5E-11	3.6E-11
Po-205	1.80 h	F	0.200	2.6E-10	0.100	2.1E-10	1.1E-10	6.6E-11	4.1E-11	3.3E-11
		M	0.200	4.0E-10	0.100	3.1E-10	1.7E-10	1.1E-10	8.1E-11	6.5E-11
		S	0.020	4.2E-10	0.010	3.2E-10	1.8E-10	1.2E-10	8.5E-11	6.9E-11
Po-207	5.83 h	F	0.200	4.8E-10	0.100	4.0E-10	2.1E-10	1.3E-10	7.3E-11	5.8E-11
		M	0.200	6.2E-10	0.100	5.1E-10	2.6E-10	1.6E-10	9.9E-11	7.8E-11
		S	0.020	6.6E-10	0.010	5.3E-10	2.7E-10	1.7E-10	1.0E-10	8.2E-11

a Dose coefficients for this element are based on age-specific biokinetic data

Table A.2.—(continued)

Nuclide	Physical half-life	Type	f_1 <1y	$e(\tau)$ 3 Months	f_1 ≥1y	$e(\tau)$ 1 Year	5 Years	10 Years	15 Years	Adult
Po-210	138 d	F	0.200	7.4E-06	0.100	4.8E-06	2.2E-06	1.3E-06	7.7E-07	6.1E-07
		M	0.200	1.5E-05	0.100	1.1E-05	6.7E-06	4.6E-06	4.0E-06	3.3E-06
		S	0.020	1.8E-05	0.010	1.4E-05	8.6E-06	5.9E-06	5.1E-06	4.3E-06
Astatine										
At-207	1.80 h	F	1.000	2.4E-09	1.000	1.7E-09	8.9E-10	5.9E-10	4.0E-10	3.3E-10
		M	1.000	9.2E-09	1.000	6.7E-09	4.3E-09	3.1E-09	2.9E-09	2.3E-09
At-211	7.21 h	F	1.000	1.4E-07	1.000	9.7E-08	4.3E-08	2.8E-08	1.7E-08	1.6E-08
		M	1.000	5.2E-07	1.000	3.7E-07	1.9E-07	1.4E-07	1.3E-07	1.1E-07
Francium										
Fr-222	0.240 h	F	1.000	9.1E-08	1.000	6.3E-08	3.0E-08	2.1E-08	1.6E-08	1.4E-08
Fr-223	0.363 h	F	1.000	1.1E-08	1.000	7.3E-09	3.2E-09	1.9E-09	1.0E-09	8.9E-10
Radium[a,b]										
Ra-223	11.4 d	F	0.600	3.0E-06	0.200	1.0E-06	4.9E-07	4.0E-07	3.3E-07	1.2E-07
		M	0.200	2.8E-05	0.100	2.1E-05	1.3E-05	9.9E-06	9.4E-06	7.4E-06
		S	0.020	3.2E-05	0.010	2.4E-05	1.5E-05	1.1E-05	1.1E-05	8.7E-06
Ra-224	3.66 d	F	0.600	1.5E-06	0.200	6.0E-07	2.9E-07	2.2E-07	1.7E-07	7.5E-08
		M	0.200	1.1E-05	0.100	8.2E-06	5.3E-06	3.9E-06	3.7E-06	3.0E-06
		S	0.020	1.2E-05	0.010	9.2E-06	5.9E-06	4.4E-06	4.2E-06	3.4E-06
Ra-225	14.8 d	F	0.600	4.0E-06	0.200	1.2E-06	5.6E-07	4.6E-07	3.8E-07	1.3E-07
		M	0.200	2.4E-05	0.100	1.8E-05	1.1E-05	8.4E-06	7.9E-06	6.3E-06
		S	0.020	2.8E-05	0.010	2.2E-05	1.4E-05	1.0E-05	9.8E-06	7.7E-06
Ra-226	1.60E+03 y	F	0.600	2.6E-06	0.200	9.4E-07	5.5E-07	7.2E-07	1.3E-06	3.6E-07
		M	0.200	1.5E-05	0.100	1.1E-05	7.0E-06	4.9E-06	4.5E-06	3.5E-06
		S	0.020	3.4E-05	0.010	2.9E-05	1.9E-05	1.2E-05	1.0E-05	9.5E-06
Ra-227	0.703 h	F	0.600	1.5E-09	0.200	1.2E-09	7.8E-10	6.1E-10	5.3E-10	4.6E-10
		M	0.200	8.0E-10	0.100	6.7E-10	4.4E-10	3.2E-10	2.9E-10	2.8E-10
		S	0.020	1.0E-09	0.010	8.5E-10	4.4E-10	2.9E-10	2.4E-10	2.2E-10
Ra-228	5.75 y	F	0.600	1.7E-05	0.200	5.7E-06	3.1E-06	3.6E-06	4.6E-06	9.0E-07
		M	0.200	1.5E-05	0.100	1.0E-05	6.3E-06	4.6E-06	4.4E-06	2.6E-06
		S	0.020	4.9E-05	0.010	4.8E-05	3.2E-05	2.0E-05	1.6E-05	1.6E-05
Actinium										
Ac-224	2.90 h	F	0.005	1.3E-07	5.0E-04	8.9E-08	4.7E-08	3.1E-08	1.4E-08	1.1E-08
		M	0.005	4.2E-07	5.0E-04	3.2E-07	2.0E-07	1.5E-07	1.4E-07	1.1E-07
		S	0.005	4.6E-07	5.0E-04	3.5E-07	2.2E-07	1.7E-07	1.6E-07	1.3E-07
Ac-225	10.0 d	F	0.005	1.1E-05	5.0E-04	7.7E-06	4.0E-06	2.6E-06	1.1E-06	8.8E-07
		M	0.005	2.8E-05	5.0E-04	2.1E-05	1.3E-05	1.0E-05	9.3E-06	7.4E-06
		S	0.005	3.1E-05	5.0E-04	2.3E-05	1.5E-05	1.1E-05	1.1E-05	8.5E-06
Ac-226	1.21 d	F	0.005	1.5E-06	5.0E-04	1.1E-06	4.0E-07	2.6E-07	1.2E-07	9.6E-08
		M	0.005	4.3E-06	5.0E-04	3.2E-06	2.1E-06	1.5E-06	1.5E-06	1.2E-06
		S	0.005	4.7E-06	5.0E-04	3.5E-06	2.3E-06	1.7E-06	1.6E-06	1.3E-06
Ac-227	21.8 y	F	0.005	1.7E-03	5.0E-04	1.6E-03	1.0E-03	7.2E-04	5.6E-04	5.5E-04
		M	0.005	5.7E-04	5.0E-04	5.5E-04	3.9E-04	2.6E-04	2.3E-04	2.2E-04
		S	0.005	2.2E-04	5.0E-04	2.0E-04	1.3E-04	8.7E-05	7.6E-05	7.2E-05
Ac-228	6.13 h	F	0.005	1.8E-07	5.0E-04	1.6E-07	9.7E-08	5.7E-08	2.9E-08	2.5E-08
		M	0.005	8.4E-08	5.0E-04	7.3E-08	4.7E-08	2.9E-08	2.0E-08	1.7E-08
		S	0.005	6.4E-08	5.0E-04	5.3E-08	3.3E-08	2.2E-08	1.9E-08	1.6E-08

a Dose coefficients for this element are based on age-specific biokinetic data

b The f_1 value for 1 to 15 year olds for Type F is 0.3

Table A.2.—(continued)

Nuclide	Physical half-life	Type	f_1 <1y	$e(\tau)$ 3 Months	f_1 ≥1y	$e(\tau)$ 1 Year	5 Years	10 Years	15 Years	Adult
Thorium[a]										
Th-226	0.515 h	F	0.005	1.4E-07	5.0E-04	1.0E-07	4.8E-08	3.4E-08	2.5E-08	2.2E-08
		M	0.005	3.0E-07	5.0E-04	2.1E-07	1.1E-07	8.3E-08	7.0E-08	5.8E-08
		S	0.005	3.1E-07	5.0E-04	2.2E-07	1.2E-07	8.8E-08	7.5E-08	6.1E-08
Th-227	18.7 d	F	0.005	8.4E-06	5.0E-04	5.2E-06	2.6E-06	1.6E-06	1.0E-06	6.7E-07
		M	0.005	3.2E-05	5.0E-04	2.5E-05	1.6E-05	1.1E-05	1.1E-05	8.5E-06
		S	0.005	3.9E-05	5.0E-04	3.0E-05	1.9E-05	1.4E-05	1.3E-05	1.0E-05
Th-228	1.91 y	F	0.005	1.8E-04	5.0E-04	1.5E-04	8.3E-05	5.2E-05	3.6E-05	2.9E-05
		M	0.005	1.3E-04	5.0E-04	1.1E-04	6.8E-05	4.6E-05	3.9E-05	3.2E-05
		S	0.005	1.6E-04	5.0E-04	1.3E-04	8.2E-05	5.5E-05	4.7E-05	4.0E-05
Th-229	7.34E+03 y	F	0.005	5.4E-04	5.0E-04	5.1E-04	3.6E-04	2.9E-04	2.4E-04	2.4E-04
		M	0.005	2.3E-04	5.0E-04	2.1E-04	1.6E-04	1.2E-04	1.1E-04	1.1E-04
		S	0.005	2.1E-04	5.0E-04	1.9E-04	1.3E-04	8.7E-05	7.6E-05	7.1E-05
Th-230	7.70E+04 y	F	0.005	2.1E-04	5.0E-04	2.0E-04	1.4E-04	1.1E-04	9.9E-05	1.0E-04
		M	0.005	7.7E-05	5.0E-04	7.4E-05	5.5E-05	4.3E-05	4.2E-05	4.3E-05
		S	0.005	4.0E-05	5.0E-04	3.5E-05	2.4E-05	1.6E-05	1.5E-05	1.4E-05
Th-231	1.06 d	F	0.005	1.1E-09	5.0E-04	7.2E-10	2.6E-10	1.6E-10	9.2E-11	7.8E-11
		M	0.005	2.2E-09	5.0E-04	1.6E-09	8.0E-10	4.8E-10	3.8E-10	3.1E-10
		S	0.005	2.4E-09	5.0E-04	1.7E-09	7.6E-10	5.2E-10	4.1E-10	3.3E-10
Th-232	1.40E+10 y	F	0.005	2.3E-04	5.0E-04	2.2E-04	1.6E-04	1.3E-04	1.2E-04	1.1E-04
		M	0.005	8.3E-05	5.0E-04	8.1E-05	6.3E-05	5.0E-05	4.7E-05	4.5E-05
		S	0.005	5.4E-05	5.0E-04	5.0E-05	3.7E-05	2.6E-05	2.5E-05	2.5E-05
Th-234	24.1 d	F	0.005	4.0E-08	5.0E-04	2.5E-08	1.1E-08	6.1E-09	3.5E-09	2.5E-09
		M	0.005	3.9E-08	5.0E-04	2.9E-08	1.5E-08	1.0E-08	7.9E-09	6.6E-09
		S	0.005	4.1E-08	5.0E-04	3.1E-08	1.7E-08	1.1E-08	9.1E-09	7.7E-09
Protactinium										
Pa-227	0.638 h	M	0.005	3.6E-07	5.0E-04	2.6E-07	1.4E-07	1.0E-07	9.0E-08	7.4E-08
		S	0.005	3.8E-07	5.0E-04	2.8E-07	1.5E-07	1.1E-07	8.1E-08	8.0E-08
Pa-228	22.0 h	M	0.005	2.6E-07	5.0E-04	2.1E-07	1.3E-07	8.8E-08	7.7E-08	6.4E-08
		S	0.005	2.9E-07	5.0E-04	2.4E-07	1.5E-07	1.0E-07	9.1E-08	7.5E-08
Pa-230	17.4 d	M	0.005	2.4E-06	5.0E-04	1.8E-06	1.1E-06	8.3E-07	7.6E-07	6.1E-07
		S	0.005	2.9E-06	5.0E-04	2.2E-06	1.4E-06	1.0E-06	9.6E-07	7.6E-07
Pa-231	3.27E+04 y	M	0.005	2.2E-04	5.0E-04	2.3E-04	1.9E-04	1.5E-04	1.5E-04	1.4E-04
		S	0.005	7.4E-05	5.0E-04	6.9E-05	5.2E-05	3.9E-05	3.6E-05	3.4E-05
Pa-232	1.31 d	M	0.005	1.9E-08	5.0E-04	1.8E-08	1.4E-08	1.1E-08	1.0E-08	1.0E-08
		S	0.005	1.0E-08	5.0E-04	8.7E-09	5.9E-09	4.1E-09	3.7E-09	3.5E-09
Pa-233	27.0 d	M	0.005	1.5E-08	5.0E-04	1.1E-08	6.5E-09	4.7E-09	4.1E-09	3.3E-09
		S	0.005	1.7E-08	5.0E-04	1.3E-08	7.5E-09	5.5E-09	4.9E-09	3.9E-09
Pa-234	6.70 h	M	0.005	2.8E-09	5.0E-04	2.0E-09	1.0E-09	6.8E-10	4.7E-10	3.8E-10
		S	0.005	2.9E-09	5.0E-04	2.1E-09	1.1E-09	7.1E-10	5.0E-10	4.0E-10
Uranium[a]										
U-230	20.8 d	F	0.040	3.2E-06	0.020	1.5E-06	7.2E-07	5.4E-07	4.1E-07	3.8E-07
		M	0.040	4.9E-05	0.020	3.7E-05	2.4E-05	1.8E-05	1.7E-05	1.3E-05
		S	0.020	5.8E-05	0.002	4.4E-05	2.8E-05	2.1E-05	2.0E-05	1.6E-05
U-231	4.20 d	F	0.040	8.9E-10	0.020	6.2E-10	3.1E-10	1.4E-10	1.0E-10	6.2E-11
		M	0.040	2.4E-09	0.020	1.7E-09	9.4E-10	5.5E-10	4.6E-10	3.8E-10
		S	0.020	2.6E-09	0.002	1.9E-09	9.0E-10	6.1E-10	4.9E-10	4.0E-10

a   Dose coefficients for this element are based on age-specific biokinetic data

Table A.2.—(continued)

Nuclide	Physical half-life	Type	f_1 <1y	$e(\tau)$ 3 Months	f_1 ≥1y	$e(\tau)$ 1 Year	5 Years	10 Years	15 Years	Adult
U-232	72.0 y	F	0.040	1.6E-05	0.020	1.0E-05	6.9E-06	6.8E-06	7.5E-06	4.0E-06
		M	0.040	3.0E-05	0.020	2.4E-05	1.6E-05	1.1E-05	1.0E-05	7.8E-06
		S	0.020	1.0E-04	0.002	9.7E-05	6.6E-05	4.3E-05	3.8E-05	3.7E-05
U-233	1.58E+05 y	F	0.040	2.2E-06	0.020	1.4E-06	9.4E-07	8.4E-07	8.6E-07	5.8E-07
		M	0.040	1.5E-05	0.020	1.1E-05	7.2E-06	4.9E-06	4.3E-06	3.6E-06
		S	0.020	3.4E-05	0.002	3.0E-05	1.9E-05	1.2E-05	1.1E-05	9.6E-06
U-234	2.44E+05 y	F	0.040	2.1E-06	0.020	1.4E-06	9.0E-07	8.0E-07	8.2E-07	5.6E-07
		M	0.040	1.5E-05	0.020	1.1E-05	7.0E-06	4.8E-06	4.2E-06	3.5E-06
		S	0.020	3.3E-05	0.002	2.9E-05	1.9E-05	1.2E-05	1.0E-05	9.4E-06
U-235	7.04E+08 y	F	0.040	2.0E-06	0.020	1.3E-06	8.5E-07	7.5E-07	7.7E-07	5.2E-07
		M	0.040	1.3E-05	0.020	1.0E-05	6.3E-06	4.3E-06	3.7E-06	3.1E-06
		S	0.020	3.0E-05	0.002	2.6E-05	1.7E-05	1.1E-05	9.2E-06	8.5E-06
U-236	2.34E+07 y	F	0.040	2.0E-06	0.020	1.3E-06	8.5E-07	7.5E-07	7.8E-07	5.3E-07
		M	0.040	1.4E-05	0.020	1.0E-05	6.5E-06	4.5E-06	3.9E-06	3.2E-06
		S	0.020	3.1E-05	0.002	2.7E-05	1.8E-05	1.1E-05	9.5E-06	8.7E-06
U-237	6.75 d	F	0.040	1.8E-09	0.020	1.5E-09	6.6E-10	4.2E-10	1.9E-10	1.8E-10
		M	0.040	7.8E-09	0.020	5.7E-09	3.3E-09	2.4E-09	2.1E-09	1.7E-09
		S	0.020	8.7E-09	0.002	6.4E-09	3.7E-09	2.7E-09	2.4E-09	1.9E-09
U-238	4.47E+09 y	F	0.040	1.9E-06	0.020	1.3E-06	8.2E-07	7.3E-07	7.4E-07	5.0E-07
		M	0.040	1.2E-05	0.020	9.4E-06	5.9E-06	4.0E-06	3.4E-06	2.9E-06
		S	0.020	2.9E-05	0.002	2.5E-05	1.6E-05	1.0E-05	8.7E-06	8.0E-06
U-239	0.392 h	F	0.040	1.0E-10	0.020	6.6E-11	2.9E-11	1.9E-11	1.2E-11	1.0E-11
		M	0.040	1.8E-10	0.020	1.2E-10	5.6E-11	3.8E-11	2.7E-11	2.2E-11
		S	0.020	1.9E-10	0.002	1.2E-10	5.9E-11	4.0E-11	2.9E-11	2.4E-11
U-240	14.1 h	F	0.040	2.4E-09	0.020	1.6E-09	7.1E-10	4.5E-10	2.3E-10	2.0E-10
		M	0.040	4.6E-09	0.020	3.1E-09	1.7E-09	1.1E-09	6.5E-10	5.3E-10
		S	0.020	4.9E-09	0.002	3.3E-09	1.6E-09	1.1E-09	7.0E-10	5.8E-10
Neptunium [a]										
Np-232	0.245 h	F	0.005	2.0E-10	5.0E-04	1.9E-10	1.2E-10	1.1E-10	1.1E-10	1.2E-10
		M	0.005	8.9E-11	5.0E-04	8.1E-11	5.5E-11	4.5E-11	4.7E-11	5.0E-11
		S	0.005	1.2E-10	5.0E-04	9.7E-11	5.8E-11	3.9E-11	2.5E-11	2.4E-11
Np-233	0.603 h	F	0.005	1.1E-11	5.0E-04	8.7E-12	4.2E-12	2.5E-12	1.4E-12	1.1E-12
		M	0.005	1.5E-11	5.0E-04	1.1E-11	5.5E-12	3.3E-12	2.1E-12	1.6E-12
		S	0.005	1.5E-11	5.0E-04	1.2E-11	5.7E-12	3.4E-12	2.1E-12	1.7E-12
Np-234	4.40 d	F	0.005	2.9E-09	5.0E-04	2.2E-09	1.1E-09	7.2E-10	4.3E-10	3.5E-10
		M	0.005	3.8E-09	5.0E-04	3.0E-09	1.6E-09	1.0E-09	6.5E-10	5.3E-10
		S	0.005	3.9E-09	5.0E-04	3.1E-09	1.6E-09	1.0E-09	6.8E-10	5.5E-10
Np-235	1.08 y	F	0.005	4.2E-09	5.0E-04	3.5E-09	1.9E-09	1.1E-09	7.5E-10	6.3E-10
		M	0.005	2.3E-09	5.0E-04	1.9E-09	1.1E-09	6.8E-10	5.1E-10	4.2E-10
		S	0.005	2.6E-09	5.0E-04	2.2E-09	1.3E-09	8.3E-10	6.3E-10	5.2E-10
Np-236	1.15E+05 y	F	0.005	8.9E-06	5.0E-04	9.1E-06	7.2E-06	7.5E-06	7.9E-06	8.0E-06
		M	0.005	3.0E-06	5.0E-04	3.1E-06	2.7E-06	2.7E-06	3.1E-06	3.2E-06
		S	0.005	1.6E-06	5.0E-04	1.6E-06	1.3E-06	1.0E-06	1.0E-06	1.0E-06
Np-236	22.5 h	F	0.005	2.8E-08	5.0E-04	2.6E-08	1.5E-08	1.1E-08	8.9E-09	9.0E-09
		M	0.005	1.6E-08	5.0E-04	1.4E-08	8.9E-09	6.2E-09	5.6E-09	5.3E-09
		S	0.005	1.6E-08	5.0E-04	1.3E-08	8.5E-09	5.7E-09	4.8E-09	4.2E-09
Np-237	.14E+06 y	F	0.005	9.8E-05	5.0E-04	9.3E-05	6.0E-05	5.0E-05	4.7E-05	5.0E-05
		M	0.005	4.4E-05	5.0E-04	4.0E-05	2.8E-05	2.2E-05	2.2E-05	2.3E-05
		S	0.005	3.7E-05	5.0E-04	3.2E-05	2.1E-05	1.4E-05	1.3E-05	1.2E-05

[a] Dose coefficients for this element are based on age-specific biokinetic data

Table A.2.—(continued)

Nuclide	Physical half-life	Type	f_1 <1y	e(τ) 3 Months	f_1 ≥1y	e(τ) 1 Year	5 Years	10 Years	15 Years	Adult
Np-238	2.12 d	F	0.005	9.0E-09	5.0E-04	7.9E-09	4.8E-09	3.7E-09	3.3E-09	3.5E-09
		M	0.005	7.3E-09	5.0E-04	5.8E-09	3.4E-09	2.5E-09	2.2E-09	2.1E-09
		S	0.005	8.1E-09	5.0E-04	6.2E-09	3.2E-09	2.1E-09	1.7E-09	1.5E-09
Np-239	2.36 d	F	0.005	2.6E-09	5.0E-04	1.4E-09	6.3E-10	3.8E-10	2.1E-10	1.7E-10
		M	0.005	5.9E-09	5.0E-04	4.2E-09	2.0E-09	1.4E-09	1.2E-09	9.3E-10
		S	0.005	5.6E-09	5.0E-04	4.0E-09	2.2E-09	1.6E-09	1.3E-09	1.0E-09
Np-240	1.08 h	F	0.005	3.6E-10	5.0E-04	2.6E-10	1.2E-10	7.7E-11	4.7E-11	4.0E-11
		M	0.005	6.3E-10	5.0E-04	4.4E-10	2.2E-10	1.4E-10	1.0E-10	8.5E-11
		S	0.005	6.5E-10	5.0E-04	4.6E-10	2.3E-10	1.5E-10	1.1E-10	9.0E-11
Plutonium[a]										
Pu-234	8.80 h	F	0.005	3.0E-08	5.0E-04	2.0E-08	9.8E-09	5.7E-09	3.6E-09	3.0E-09
		M	0.005	7.8E-08	5.0E-04	5.9E-08	3.7E-08	2.8E-08	2.6E-08	2.1E-08
		S	1.0E-04	8.7E-08	1.0E-05	6.6E-08	4.2E-08	3.1E-08	3.0E-08	2.4E-08
Pu-235	0.422 h	F	0.005	1.0E-11	5.0E-04	7.9E-12	3.9E-12	2.2E-12	1.3E-12	1.0E-12
		M	0.005	1.3E-11	5.0E-04	1.0E-11	5.0E-12	2.9E-12	1.9E-12	1.4E-12
		S	1.0E-04	1.3E-11	1.0E-05	1.0E-11	5.1E-12	3.0E-12	1.9E-12	1.5E-12
Pu-236	2.85 y	F	0.005	1.0E-04	5.0E-04	9.5E-05	6.1E-05	4.4E-05	3.7E-05	4.0E-05
		M	0.005	4.8E-05	5.0E-04	4.3E-05	2.9E-05	2.1E-05	1.9E-05	2.0E-05
		S	1.0E-04	3.6E-05	1.0E-05	3.1E-05	2.0E-05	1.4E-05	1.2E-05	1.0E-05
Pu-237	45.3 d	F	0.005	2.2E-09	5.0E-04	1.6E-09	7.9E-10	4.8E-10	2.9E-10	2.6E-10
		M	0.005	1.9E-09	5.0E-04	1.4E-09	8.2E-10	5.4E-10	4.3E-10	3.5E-10
		S	1.0E-04	2.0E-09	1.0E-05	1.5E-09	8.8E-10	5.9E-10	4.8E-10	3.9E-10
Pu-238	87.7 y	F	0.005	2.0E-04	5.0E-04	1.9E-04	1.4E-04	1.1E-04	1.0E-04	1.1E-04
		M	0.005	7.8E-05	5.0E-04	7.4E-05	5.6E-05	4.4E-05	4.3E-05	4.6E-05
		S	1.0E-04	4.5E-05	1.0E-05	4.0E-05	2.7E-05	1.9E-05	1.7E-05	1.6E-05
Pu-239	2.41E+04 y	F	0.005	2.1E-04	5.0E-04	2.0E-04	1.5E-04	1.2E-04	1.1E-04	1.2E-04
		M	0.005	8.0E-05	5.0E-04	7.7E-05	6.0E-05	4.8E-05	4.7E-05	5.0E-05
		S	1.0E-04	4.3E-05	1.0E-05	3.9E-05	2.7E-05	1.9E-05	1.7E-05	1.6E-05
Pu-240	6.54E+03 y	F	0.005	2.1E-04	5.0E-04	2.0E-04	1.5E-04	1.2E-04	1.1E-04	1.2E-04
		M	0.005	8.0E-05	5.0E-04	7.7E-05	6.0E-05	4.8E-05	4.7E-05	5.0E-05
		S	1.0E-04	4.3E-05	1.0E-05	3.9E-05	2.7E-05	1.9E-05	1.7E-05	1.6E-05
Pu-241	14.4 y	F	0.005	2.8E-06	5.0E-04	2.9E-06	2.6E-06	2.4E-06	2.2E-06	2.3E-06
		M	0.005	9.1E-07	5.0E-04	9.7E-07	9.2E-07	8.3E-07	8.6E-07	9.0E-07
		S	1.0E-04	2.2E-07	1.0E-05	2.3E-07	2.0E-07	1.7E-07	1.7E-07	1.7E-07
Pu-242	3.76E+05 y	F	0.005	2.0E-04	5.0E-04	1.9E-04	1.4E-04	1.2E-04	1.1E-04	1.1E-04
		M	0.005	7.6E-05	5.0E-04	7.3E-05	5.7E-05	4.5E-05	4.5E-05	4.8E-05
		S	1.0E-04	4.0E-05	1.0E-05	3.6E-05	2.5E-05	1.7E-05	1.6E-05	1.5E-05
Pu-243	4.95 h	F	0.005	2.7E-10	5.0E-04	1.9E-10	8.8E-11	5.7E-11	3.5E-11	3.2E-11
		M	0.005	5.6E-10	5.0E-04	3.9E-10	1.9E-10	1.3E-10	8.7E-11	8.3E-11
		S	1.0E-04	6.0E-10	1.0E-05	4.1E-10	2.0E-10	1.4E-10	9.2E-11	8.6E-11
Pu-244	8.26E+07 y	F	0.005	2.0E-04	5.0E-04	1.9E-04	1.4E-04	1.2E-04	1.1E-04	1.1E-04
		M	0.005	7.4E-05	5.0E-04	7.2E-05	5.6E-05	4.5E-05	4.4E-05	4.7E-05
		S	1.0E-04	3.9E-05	1.0E-05	3.5E-05	2.4E-05	1.7E-05	1.5E-05	1.5E-05
Pu-245	10.5 h	F	0.005	1.8E-09	5.0E-04	1.3E-09	5.6E-10	3.5E-10	1.9E-10	1.6E-10
		M	0.005	3.6E-09	5.0E-04	2.5E-09	1.2E-09	8.0E-10	5.0E-10	4.0E-10
		S	1.0E-04	3.8E-09	1.0E-05	2.6E-09	1.3E-09	8.5E-10	5.4E-10	4.3E-10
Pu-246	10.9 d	F	0.005	2.0E-08	5.0E-04	1.4E-08	7.0E-09	4.4E-09	2.8E-09	2.5E-09
		M	0.005	3.5E-08	5.0E-04	2.6E-08	1.5E-08	1.1E-08	9.1E-09	7.4E-09
		S	1.0E-04	3.8E-08	1.0E-05	2.8E-08	1.6E-08	1.2E-08	1.0E-08	8.0E-09

a Dose coefficients for this element are based on age-specific biokinetic data

Table A.2.—(continued)

Nuclide	Physical half-life	Type	f_1 <1y	$e(\tau)$ 3 Months	f_1 ≥1y	$e(\tau)$ 1 Year	5 Years	10 Years	15 Years	Adult
Americium[a]										
Am-237	1.22 h	F	0.005	9.8E-11	5.0E-04	7.3E-11	3.5E-11	2.2E-11	1.3E-11	1.1E-11
		M	0.005	1.7E-10	5.0E-04	1.2E-10	6.2E-11	4.1E-11	3.0E-11	2.5E-11
		S	0.005	1.7E-10	5.0E-04	1.3E-10	6.5E-11	4.3E-11	3.2E-11	2.6E-11
Am-238	1.63 h	F	0.005	4.1E-10	5.0E-04	3.8E-10	2.5E-10	2.0E-10	1.8E-10	1.9E-10
		M	0.005	3.1E-10	5.0E-04	2.6E-10	1.3E-10	9.6E-11	8.8E-11	9.0E-11
		S	0.005	2.7E-10	5.0E-04	2.2E-10	1.3E-10	8.2E-11	6.1E-11	5.4E-11
Am-239	11.9 h	F	0.005	8.1E-10	5.0E-04	5.8E-10	2.6E-10	1.6E-10	9.1E-11	7.6E-11
		M	0.005	1.5E-09	5.0E-04	1.1E-09	5.6E-10	3.7E-10	2.7E-10	2.2E-10
		S	0.005	1.6E-09	5.0E-04	1.1E-09	5.9E-10	4.0E-10	2.5E-10	2.4E-10
Am-240	2.12 d	F	0.005	2.0E-09	5.0E-04	1.7E-09	8.8E-10	5.7E-10	3.6E-10	2.3E-10
		M	0.005	2.9E-09	5.0E-04	2.2E-09	1.2E-09	7.7E-10	5.3E-10	4.3E-10
		S	0.005	3.0E-09	5.0E-04	2.3E-09	1.2E-09	7.8E-10	5.3E-10	4.3E-10
Am-241	4.32E+02 y	F	0.005	1.8E-04	5.0E-04	1.8E-04	1.2E-04	1.0E-04	9.2E-05	9.6E-05
		M	0.005	7.3E-05	5.0E-04	6.9E-05	5.1E-05	4.0E-05	4.0E-05	4.2E-05
		S	0.005	4.6E-05	5.0E-04	4.0E-05	2.7E-05	1.9E-05	1.7E-05	1.6E-05
Am-242	16.0 h	F	0.005	9.2E-08	5.0E-04	7.1E-08	3.5E-08	2.1E-08	1.4E-08	1.1E-08
		M	0.005	7.6E-08	5.0E-04	5.9E-08	3.6E-08	2.4E-08	2.1E-08	1.7E-08
		S	0.005	8.0E-08	5.0E-04	6.2E-08	3.9E-08	2.7E-08	2.4E-08	2.0E-08
Am-242m	1.52E+02 y	F	0.005	1.6E-04	5.0E-04	1.5E-04	1.1E-04	9.4E-05	8.8E-05	9.2E-05
		M	0.005	5.2E-05	5.0E-04	5.3E-05	4.1E-05	3.4E-05	3.5E-05	3.7E-05
		S	0.005	2.5E-05	5.0E-04	2.4E-05	1.7E-05	1.2E-05	1.1E-05	1.1E-05
Am-243	7.38E+03 y	F	0.005	1.8E-04	5.0E-04	1.7E-04	1.2E-04	1.0E-04	9.1E-05	9.6E-05
		M	0.005	7.2E-05	5.0E-04	6.8E-05	5.0E-05	4.0E-05	4.0E-05	4.1E-05
		S	0.005	4.4E-05	5.0E-04	3.9E-05	2.6E-05	1.8E-05	1.6E-05	1.5E-05
Am-244	10.1 h	F	0.005	1.0E-08	5.0E-04	9.2E-09	5.6E-09	4.1E-09	3.5E-09	3.7E-09
		M	0.005	6.0E-09	5.0E-04	5.0E-09	3.2E-09	2.2E-09	2.0E-09	2.0E-09
		S	0.005	6.1E-09	5.0E-04	4.8E-09	2.4E-09	1.6E-09	1.4E-09	1.2E-09
Am-244m	0.433 h	F	0.005	4.6E-10	5.0E-04	4.0E-10	2.4E-10	1.8E-10	1.5E-10	1.6E-10
		M	0.005	3.3E-10	5.0E-04	2.1E-10	1.3E-10	9.2E-11	8.3E-11	8.4E-11
		S	0.005	3.0E-10	5.0E-04	2.2E-10	1.2E-10	8.1E-11	5.5E-11	5.7E-11
Am-245	2.05 h	F	0.005	2.1E-10	5.0E-04	1.4E-10	6.2E-11	4.0E-11	2.4E-11	2.1E-11
		M	0.005	3.9E-10	5.0E-04	2.6E-10	1.3E-10	8.7E-11	6.4E-11	5.3E-11
		S	0.005	4.1E-10	5.0E-04	2.8E-10	1.3E-10	9.2E-11	6.8E-11	5.6E-11
Am-246	0.650 h	F	0.005	3.0E-10	5.0E-04	2.0E-10	9.3E-11	6.1E-11	3.8E-11	3.3E-11
		M	0.005	5.0E-10	5.0E-04	3.4E-10	1.6E-10	1.1E-10	7.9E-11	6.6E-11
		S	0.005	5.3E-10	5.0E-04	3.6E-10	1.7E-10	1.2E-10	8.3E-11	6.9E-11
Am-246m	0.417 h	F	0.005	1.3E-10	5.0E-04	8.9E-11	4.2E-11	2.6E-11	1.6E-11	1.4E-11
		M	0.005	1.9E-10	5.0E-04	1.3E-10	6.1E-11	4.0E-11	2.6E-11	2.2E-11
		S	0.005	2.0E-10	5.0E-04	1.4E-10	6.4E-11	4.1E-11	2.7E-11	2.3E-11
Curium[a]										
Cm-238	2.40 h	F	0.005	7.7E-09	5.0E-04	5.4E-09	2.6E-09	1.8E-09	9.2E-10	7.8E-10
		M	0.005	2.1E-08	5.0E-04	1.5E-08	7.9E-09	5.9E-09	5.6E-09	4.5E-09
		S	0.005	2.2E-08	5.0E-04	1.6E-08	8.6E-09	6.4E-09	6.1E-09	4.9E-09
Cm-240	27.0 d	F	0.005	8.3E-06	5.0E-04	6.3E-06	3.2E-06	2.0E-06	1.5E-06	1.3E-06
		M	0.005	1.2E-05	5.0E-04	9.1E-06	5.8E-06	4.2E-06	3.8E-06	3.2E-06
		S	0.005	1.3E-05	5.0E-04	9.9E-06	6.4E-06	4.6E-06	4.3E-06	3.5E-06

a Dose coefficients for this element are based on age-specific biokinetic data

Table A.2.—(continued)

Nuclide	Physical half-life	Type	f_1 <1y	$e(\tau)$ 3 Months	f_1 ≥1y	$e(\tau)$ 1 Year	5 Years	10 Years	15 Years	Adult
Cm-241	32.8 d	F	0.005	1.1E-07	5.0E-04	8.9E-08	4.9E-08	3.5E-08	2.8E-08	2.7E-08
		M	0.005	1.3E-07	5.0E-04	1.0E-07	6.6E-08	4.8E-08	4.4E-08	3.7E-08
		S	0.005	1.4E-07	5.0E-04	1.1E-07	6.9E-08	4.9E-08	4.5E-08	3.7E-08
Cm-242	163 d	F	0.005	2.7E-05	5.0E-04	2.1E-05	1.0E-05	6.1E-06	4.0E-06	3.3E-06
		M	0.005	2.2E-05	5.0E-04	1.8E-05	1.1E-05	7.3E-06	6.4E-06	5.2E-06
		S	0.005	2.4E-05	5.0E-04	1.9E-05	1.2E-05	8.2E-06	7.3E-06	5.9E-06
Cm-243	28.5 y	F	0.005	1.6E-04	5.0E-04	1.5E-04	9.5E-05	7.3E-05	6.5E-05	6.9E-05
		M	0.005	6.7E-05	5.0E-04	6.1E-05	4.2E-05	3.1E-05	3.0E-05	3.1E-05
		S	0.005	4.6E-05	5.0E-04	4.0E-05	2.6E-05	1.8E-05	1.6E-05	1.5E-05
Cm-244	18.1 y	F	0.005	1.5E-04	5.0E-04	1.3E-04	8.3E-05	6.1E-05	5.3E-05	5.7E-05
		M	0.005	6.2E-05	5.0E-04	5.7E-05	3.7E-05	2.7E-05	2.6E-05	2.7E-05
		S	0.005	4.4E-05	5.0E-04	3.8E-05	2.5E-05	1.7E-05	1.5E-05	1.3E-05
Cm-245	8.50E+03 y	F	0.005	1.9E-04	5.0E-04	1.8E-04	1.2E-04	1.0E-04	9.4E-05	9.9E-05
		M	0.005	7.3E-05	5.0E-04	6.9E-05	5.1E-05	4.1E-05	4.1E-05	4.2E-05
		S	0.005	4.5E-05	5.0E-04	4.0E-05	2.7E-05	1.9E-05	1.7E-05	1.6E-05
Cm-246	4.73E+03 y	F	0.005	1.9E-04	5.0E-04	1.8E-04	1.2E-04	1.0E-04	9.4E-05	9.8E-05
		M	0.005	7.3E-05	5.0E-04	6.9E-05	5.1E-05	4.1E-05	4.1E-05	4.2E-05
		S	0.005	4.6E-05	5.0E-04	4.0E-05	2.7E-05	1.9E-05	1.7E-05	1.6E-05
Cm-247	1.56E+07 y	F	0.005	1.7E-04	5.0E-04	1.6E-04	1.1E-04	9.4E-05	8.6E-05	9.0E-05
		M	0.005	6.7E-05	5.0E-04	6.3E-05	4.7E-05	3.7E-05	3.7E-05	3.9E-05
		S	0.005	4.1E-05	5.0E-04	3.6E-05	2.4E-05	1.7E-05	1.5E-05	1.4E-05
Cm-248	3.39E+05 y	F	0.005	6.8E-04	5.0E-04	6.5E-04	4.5E-04	3.7E-04	3.4E-04	3.6E-04
		M	0.005	2.5E-04	5.0E-04	2.4E-04	1.8E-04	1.4E-04	1.4E-04	1.5E-04
		S	0.005	1.4E-04	5.0E-04	1.2E-04	8.2E-05	5.6E-05	5.0E-05	4.8E-05
Cm-249	1.07 h	F	0.005	1.8E-10	5.0E-04	9.8E-11	5.9E-11	4.6E-11	4.0E-11	4.0E-11
		M	0.005	2.4E-10	5.0E-04	1.6E-10	8.2E-11	5.8E-11	3.7E-11	3.3E-11
		S	0.005	2.4E-10	5.0E-04	1.6E-10	7.8E-11	5.3E-11	3.9E-11	3.3E-11
Cm-250	6.90E+03 y	F	0.005	3.9E-03	5.0E-04	3.7E-03	2.6E-03	2.1E-03	2.0E-03	2.1E-03
		M	0.005	1.4E-03	5.0E-04	1.3E-03	9.9E-04	7.9E-04	7.9E-04	8.4E-04
		S	0.005	7.2E-04	5.0E-04	6.5E-04	4.4E-04	3.0E-04	2.7E-04	2.6E-04
Berkelium										
Bk-245	4.94 d	M	0.005	8.8E-09	5.0E-04	6.6E-09	4.0E-09	2.9E-09	2.6E-09	2.1E-09
Bk-246	1.83 d	M	0.005	2.1E-09	5.0E-04	1.7E-09	9.3E-10	6.0E-10	4.0E-10	3.3E-10
Bk-247	1.38E+03 y	M	0.005	1.5E-04	5.0E-04	1.5E-04	1.1E-04	7.9E-05	7.2E-05	6.9E-05
Bk-249	320 d	M	0.005	3.3E-07	5.0E-04	3.3E-07	2.4E-07	1.8E-07	1.6E-07	1.6E-07
Bk-250	3.22 h	M	0.005	3.4E-09	5.0E-04	3.1E-09	2.0E-09	1.3E-09	1.1E-09	1.0E-09
Californium										
Cf-244	0.323 h	M	0.005	7.6E-08	5.0E-04	5.4E-08	2.8E-08	2.0E-08	1.6E-08	1.4E-08
Cf-246	1.49 d	M	0.005	1.7E-06	5.0E-04	1.3E-06	8.3E-07	6.1E-07	5.7E-07	4.5E-07
Cf-248	334 d	M	0.005	3.8E-05	5.0E-04	3.2E-05	2.1E-05	1.4E-05	1.0E-05	8.8E-06
Cf-249	3.50E+02 y	M	0.005	1.6E-04	5.0E-04	1.5E-04	1.1E-04	8.0E-05	.2E-05	7.0E-05
Cf-250	13.1 y	M	0.005	1.1E-04	5.0E-04	9.8E-05	6.6E-05	4.2E-05	3.5E-05	3.4E-05
Cf-251	8.98E+02 y	M	0.005	1.6E-04	5.0E-04	1.5E-04	1.1E-04	8.1E-05	7.3E-05	7.1E-05
Cf-252	2.64 y	M	0.005	9.7E-05	5.0E-04	8.7E-05	5.6E-05	3.2E-05	2.2E-05	2.0E-05

Table A.2.—(continued)

Nuclide	Physical half-life	Type	f_1 <1y	$e(\tau)$ 3 Months	f_1 ≥1y	$e(\tau)$ 1 Year	5 Years	10 Years	15 Years	Adult
Cf-253	17.8 d	M	0.005	5.4E-06	5.0E-04	4.2E-06	2.6E-06	1.9E-06	1.7E-06	1.3E-06
Cf-254	60.5 d	M	0.005	2.5E-04	5.0E-04	1.9E-04	1.1E-04	7.0E-05	4.8E-05	4.1E-05
Einsteinium										
Es-250	2.10 h	M	0.005	2.0E-09	5.0E-04	1.8E-09	1.2E-09	7.8E-10	6.4E-10	6.3E-10
Es-251	1.38 d	M	0.005	7.9E-09	5.0E-04	6.0E-09	3.9E-09	2.8E-09	2.6E-09	2.1E-09
Es-253	20.5 d	M	0.005	1.1E-05	5.0E-04	8.0E-06	5.1E-06	3.7E-06	3.4E-06	2.7E-06
Es-254	276 d	M	0.005	3.7E-05	5.0E-04	3.1E-05	2.0E-05	1.3E-05	1.0E-05	9.6E-06
Es-254m	1.64 d	M	0.005	1.7E-06	5.0E-04	1.3E-06	8.4E-07	6.3E-07	5.9E-07	4.7E-07
Fermium										
Fm-252	22.7 h	M	0.005	1.2E-06	5.0E-04	9.0E-07	5.8E-07	4.3E-07	4.0E-07	3.2E-07
Fm-253	3.00 d	M	0.005	1.5E-06	5.0E-04	1.2E-06	7.3E-07	5.4E-07	5.0E-07	4.0E-07
Fm-254	3.24 h	M	0.005	3.2E-07	5.0E-04	2.3E-07	1.3E-07	9.8E-08	7.6E-08	6.1E-08
Fm-255	20.1 h	M	0.005	1.2E-06	5.0E-04	7.3E-07	4.7E-07	3.5E-07	3.4E-07	2.7E-07
Fm-257	101 d	M	0.005	3.3E-05	5.0E-04	2.6E-05	1.6E-05	1.1E-05	8.8E-06	7.1E-06
Mendelevium										
Md-257	5.20 h	M	0.005	1.0E-07	5.0E-04	8.2E-08	5.1E-08	3.6E-08	3.1E-08	2.5E-08
Md-258	55.0 d	M	0.005	2.4E-05	5.0E-04	1.9E-05	1.2E-05	8.6E-06	7.3E-06	5.9E-06

Table A.3. Inhalation dose coefficients, $e(\tau)$, to age 70 y (Sv Bq^{-1}) for soluble or reactive gases and vapours (Classes SR-1 and SR-2)

Nuclide	Physical half-life	Absorption deposit	f_1 <1y	$e(\tau)$ 3 Months	f_1 ≥1y	$e(\tau)$ 1 Year	5 Years	10 Years	15 Years	Adult
Tritiated water	12.3 y	v	100	1.000	6.4E-11	1.000	4.8E-11	3.1E-11	2.3E-11	1.8E-11	1.8E-11
Elemental hydrogen	12.3 y	v	0.01	1.000	6.4E-15	1.000	4.8E-15	3.1E-15	2.3E-15	1.8E-15	1.8E-15
Tritiated methane	12.3 y	v	1	1.000	6.4E-13	1.000	4.8E-13	3.1E-13	2.3E-13	1.8E-13	1.8E-13
Organically bound tritium	12.3 y	v	100	1.000	1.1E-10	1.000	1.1E-10	7.0E-11	5.5E-11	4.1E-11	4.1E-11
Carbon-11 vapour	0.340 h	v	100	1.000	2.8E-11	1.000	1.8E-11	9.7E-12	6.1E-12	3.8E-12	3.2E-12
Carbon-11 dioxide	0.340 h	v	100	1.000	1.8E-11	1.000	1.2E-11	6.5E-12	4.1E-12	2.5E-12	2.2E-12
Carbon-11 monoxide	0.340 h	v	40	1.000	1.0E-11	1.000	6.7E-12	3.5E-12	2.2E-12	1.4E-12	1.2E-12
Carbon-14 vapour	5.73E+03 y	v	100	1.000	1.3E-09	1.000	1.6E-09	9.7E-10	7.9E-10	5.7E-10	5.8E-10
Carbon-14 dioxide	5.73E+03 y	v	100	1.000	1.9E-11	1.000	1.9E-11	1.1E-11	8.9E-12	6.3E-12	6.2E-12
Carbon-14 monoxide	5.73E+03 y	v	40	1.000	9.1E-12	1.000	5.7E-12	2.8E-12	1.7E-12	9.9E-13	8.0E-13
Carbon disulphide-35	87.4 d	F	100	1.000	6.9E-09	0.800	4.8E-09	2.4E-09	1.4E-09	8.6E-10	7.0E-10
Sulphur-35 dioxide	87.4 d	F	85	1.000	9.4E-10	0.800	6.6E-10	3.4E-10	2.1E-10	1.3E-10	1.1E-10
Nickel-56 carbonyl	6.10 d	a	100	1.000	6.8E-09	1.000	5.2E-09	3.2E-09	2.1E-09	1.4E-09	1.2E-09
Nickel-57 carbonyl	1.50 d	a	100	1.000	3.1E-09	1.000	2.3E-09	1.4E-09	9.2E-10	6.5E-10	5.6E-10
Nickel-59 carbonyl	7.50E+04 y	a	100	1.000	4.0E-09	1.000	3.3E-09	2.0E-09	1.3E-09	9.1E-10	8.3E-10
Nickel-63 carbonyl	96.0 y	a	100	1.000	9.5E-09	1.000	8.0E-09	4.8E-09	3.0E-09	2.2E-09	2.0E-09
Nickel-65 carbonyl	2.52 h	a	100	1.000	2.0E-09	1.000	1.4E-09	8.1E-10	5.6E-10	4.0E-10	3.6E-10

a Deposition 30%:10%:20%:40% (extrathoracic:bronchial:bronchiolar:alveolar-interstitial), 0.1 day retention half-time (ICRP 1994b)

Table A.3.—(continued)

Nuclide	Physical half-life	Absorption & deposit %	f_1 <1y	$e(\tau)$ 3 Months	f_1 ≥1y	1 Year	5 Years	10 Years	15 Years	Adult
Nickel-66 carbonyl	2.27 d	a 100	1.000	1.0E-08	1.000	7.1E-09	4.0E-09	2.7E-09	1.8E-09	1.6E-09
Ruthenium-94 tetroxide	0.863 h	F 100	0.100	5.5E-10	0.050	3.5E-10	1.8E-10	1.1E-10	7.0E-11	5.6E-11
Ruthenium-97 tetroxide	2.90 d	F 100	0.100	8.7E-10	0.050	6.2E-10	3.4E-10	2.2E-10	1.4E-10	1.2E-10
Ruthenium-103 tetroxide	39.3 d	F 100	0.100	9.0E-09	0.050	6.2E-09	3.3E-09	2.1E-09	1.3E-09	1.1E-09
Ruthenium-105 tetroxide	4.44 h	F 100	0.100	1.6E-09	0.050	1.0E-09	5.3E-10	3.2E-10	2.2E-10	1.8E-10
Ruthenium-106 tetroxide	1.01 y	F 100	0.100	1.6E-07	0.050	1.1E-07	6.1E-08	3.7E-08	2.2E-08	1.8E-08
Tellurium-116 vapour	2.49 h	F 100	0.600	5.9E-10	0.300	4.4E-10	2.5E-10	1.6E-10	1.1E-10	8.7E-11
Tellurium-121 vapour	17.0 d	F 100	0.600	3.0E-09	0.300	2.4E-09	1.4E-09	9.6E-10	6.7E-10	5.1E-10
Tellurium-121m vapour	154 d	F 100	0.600	3.5E-08	0.300	2.7E-08	1.6E-08	9.8E-09	6.6E-09	5.5E-09
Tellurium-123 vapour	1.00E+13 y	F 100	0.600	2.8E-08	0.300	2.5E-08	1.9E-08	1.5E-08	1.3E-08	1.2E-08
Tellurium-123m vapour	120 d	F 100	0.600	2.5E-08	0.300	1.8E-08	1.0E-08	5.7E-09	3.5E-09	2.9E-09
Tellurium-125m vapour	58.0 d	F 100	0.600	1.5E-08	0.300	1.1E-08	5.9E-09	3.2E-09	1.9E-09	1.5E-09
Tellurium-127 vapour	9.35 h	F 100	0.600	6.1E-10	0.300	4.4E-10	2.3E-10	1.4E-10	9.2E-11	7.7E-11
Tellurium-127m vapour	109 d	F 100	0.600	5.3E-08	0.300	3.7E-08	1.9E-08	1.0E-08	6.1E-09	4.6E-09
Tellurium-129 vapour	1.16 h	F 100	0.600	2.5E-10	0.300	1.7E-10	9.4E-11	6.2E-11	4.3E-11	3.7E-11
Tellurium-129m vapour	33.6 d	F 100	0.600	4.8E-08	0.300	3.2E-08	1.6E-08	8.5E-09	5.1E-09	3.7E-09
Tellurium-131 vapour	0.417 h	F 100	0.600	5.1E-10	0.300	4.5E-10	2.6E-10	1.4E-10	9.5E-11	6.8E-11
Tellurium-131m vapour	1.25 d	F 100	0.600	2.1E-08	0.300	1.9E-08	1.1E-08	5.6E-09	3.7E-09	2.4E-09

Table A.3.—(continued)

Nuclide	Physical half-life	Absorption % deposit	f_1 <1y	e(τ) 3 Months	f_1 ≥1y	e(τ) 1 Year	5 Years	10 Years	15 Years	Adult
Tellurium-132 vapour	3.26 d	F 100	0.600	5.4E-08	0.300	4.5E-08	2.4E-08	1.2E-08	7.6E-09	5.1E-09
Tellurium-133 vapour	0.207 h	F 100	0.600	5.5E-10	0.300	4.7E-10	2.5E-10	1.2E-10	8.1E-11	5.6E-11
Tellurium-133m vapour	0.923 h	F 100	0.600	2.3E-09	0.300	2.0E-09	1.1E-09	5.0E-10	3.3E-10	2.2E-10
Tellurium-134 vapour	0.696 h	F 100	0.600	6.8E-10	0.300	5.5E-10	3.0E-10	1.6E-10	1.1E-10	8.4E-11
Elemental iodine-120	1.35 h	V 100	1.000	3.0E-09	1.000	2.4E-09	1.3E-09	6.4E-10	4.3E-10	3.0E-10
Elemental iodine-120m	0.883 h	V 100	1.000	1.5E-09	1.000	1.2E-09	6.4E-10	3.4E-10	2.3E-10	1.8E-10
Elemental iodine-121	2.12 h	V 100	1.000	5.7E-10	1.000	5.1E-10	3.0E-10	1.7E-10	1.2E-10	8.6E-11
Elemental iodine-123	13.2 h	V 100	1.000	2.1E-09	1.000	1.8E-09	1.0E-09	4.7E-10	3.2E-10	2.1E-10
Elemental iodine-124	4.18 d	V 100	1.000	1.1E-07	1.000	1.0E-07	5.8E-08	2.8E-08	1.8E-08	1.2E-08
Elemental iodine-125	60.1 d	V 100	1.000	4.7E-08	1.000	5.2E-08	3.7E-08	2.8E-08	2.0E-08	1.4E-08
Elemental iodine-126	13.0 d	V 100	1.000	1.9E-07	1.000	1.9E-07	1.1E-07	6.2E-08	4.1E-08	2.6E-08
Elemental iodine-128	0.416 h	V 100	1.000	4.2E-10	1.000	2.8E-10	1.6E-10	1.0E-10	7.5E-11	6.5E-11
Elemental iodine-129	1.57E+07 y	V 100	1.000	1.7E-07	1.000	2.0E-07	1.6E-07	1.7E-07	1.3E-07	9.6E-08
Elemental iodine-130	12.4 h	V 100	1.000	1.9E-08	1.000	1.7E-08	9.2E-09	4.3E-09	2.8E-09	1.9E-09
Elemental iodine-131	8.04 d	V 100	1.000	1.7E-07	1.000	1.6E-07	9.4E-08	4.8E-08	3.1E-08	2.0E-08
Elemental iodine-132	2.30 h	V 100	1.000	2.8E-09	1.000	2.3E-09	1.3E-09	6.4E-10	4.3E-10	3.1E-10
Elemental iodine-132m	1.39 h	V 100	1.000	2.4E-09	1.000	2.1E-09	1.1E-09	5.6E-10	3.8E-10	2.7E-10
Elemental iodine-133	20.8 h	V 100	1.000	4.5E-08	1.000	4.1E-08	2.1E-08	9.7E-09	6.3E-09	4.0E-09

Table A.3.—(continued)

Nuclide	Physical half-life	Absorption % deposit	f_1 <1y	$e(\tau)$ 3 Months	f_1 ≥1y	$e(\tau)$ 1 Year	5 Years	10 Years	15 Years	Adult
Elemental iodine-134	0.876 h	V 100	1.000	8.7E-10	1.000	6.9E-10	3.9E-10	2.2E-10	1.6E-10	1.5E-10
Elemental iodine-135	6.61 h	V 100	1.000	9.7E-09	1.000	8.5E-09	4.5E-09	2.1E-09	1.4E-09	9.2E-10
Methyl iodide-120	1.35 h	V 70	1.000	2.3E-09	1.000	1.9E-09	1.0E-09	4.8E-10	3.1E-10	2.0E-10
Methyl iodide-120m	0.883 h	V 70	1.000	1.0E-09	1.000	8.7E-10	4.6E-10	2.2E-10	1.5E-10	1.0E-10
Methyl iodide-121	2.12 h	V 70	1.000	4.2E-10	1.000	3.8E-10	2.2E-10	1.2E-10	8.3E-11	5.6E-11
Methyl iodide-123	13.2 h	V 70	1.000	1.6E-09	1.000	1.4E-09	7.7E-10	3.6E-10	2.4E-10	1.5E-10
Methyl iodide-124	4.18 d	V 70	1.000	8.5E-08	1.000	8.0E-08	4.5E-08	2.2E-08	1.4E-08	9.2E-09
Methyl iodide-125	60.1 d	V 70	1.000	3.7E-08	1.000	4.0E-08	2.9E-08	2.2E-08	1.6E-08	1.1E-08
Methyl iodide-126	13.0 d	V 70	1.000	1.5E-07	1.000	1.5E-07	9.0E-08	4.8E-08	3.2E-08	2.0E-08
Methyl iodide-128	0.416 h	V 70	1.000	1.5E-10	1.000	1.2E-10	6.3E-11	3.0E-11	1.9E-11	1.3E-11
Methyl iodide-129	1.57E+07 y	V 70	1.000	1.3E-07	1.000	1.5E-07	1.2E-07	1.3E-07	9.9E-08	7.4E-08
Methyl iodide-130	12.4 h	V 70	1.000	1.5E-08	1.000	1.3E-08	7.2E-09	3.3E-09	2.2E-09	1.4E-09
Methyl iodide-131	8.04 d	V 70	1.000	1.3E-07	1.000	1.3E-07	7.4E-08	3.7E-08	2.4E-08	1.5E-08
Methyl iodide-132	2.30 h	V 70	1.000	2.0E-09	1.000	1.8E-09	9.5E-10	4.4E-10	2.9E-10	1.9E-10
Methyl iodide-132m	1.39 h	V 70	1.000	1.8E-09	1.000	1.6E-09	8.3E-10	3.9E-10	2.5E-10	1.6E-10
Methyl iodide-133	20.8 h	V 70	1.000	3.5E-08	1.000	3.2E-08	1.7E-08	7.6E-09	4.9E-09	3.1E-09
Methyl iodide-134	0.876 h	V 70	1.000	5.1E-10	1.000	4.3E-10	2.3E-10	1.1E-10	7.4E-11	5.0E-11
Methyl iodide-135	6.61 h	V 70	1.000	7.5E-09	1.000	6.7E-09	3.5E-09	1.6E-09	1.1E-09	6.8E-10

Table A.3.—(continued)

Nuclide	Physical half-life	Absorption % deposit	f_1 <1y	$e(\tau)$ 3 Months	f_1 ≥1y	$e(\tau)$ 1 Year	5 Years	10 Years	15 Years	Adult
Mercury-193 vapour	3.50 h	a 70	1.000	4.2E-09	1.000	3.4E-09	2.2E-09	1.6E-09	1.2E-09	1.1E-09
Mercury-193m vapour	11.1 h	a 70	1.000	1.2E-08	1.000	9.4E-09	6.1E-09	4.5E-09	3.4E-09	3.1E-09
Mercury-194 vapour	2.60E+02 y	a 70	1.000	9.4E-08	1.000	8.3E-08	6.2E-08	5.0E-08	4.3E-08	4.0E-08
Mercury-195 vapour	9.90 h	a 70	1.000	5.3E-09	1.000	4.3E-09	2.8E-09	2.1E-09	1.6E-09	1.4E-09
Mercury-195m vapour	1.73 d	a 70	1.000	3.0E-08	1.000	2.5E-08	1.6E-08	1.2E-08	8.8E-09	8.2E-09
Mercury-197 vapour	2.67 d	a 70	1.000	1.6E-08	1.000	1.3E-08	8.4E-09	6.3E-09	4.7E-09	4.4E-09
Mercury-197m vapour	23.8 h	a 70	1.000	2.1E-08	1.000	1.7E-08	1.1E-08	8.2E-09	6.2E-09	5.8E-09
Mercury-199m vapour	0.710 h	a 70	1.000	6.5E-10	1.000	5.3E-10	3.4E-10	2.5E-10	1.9E-10	1.8E-10
Mercury-203 vapour	46.6 d	a 70	1.000	3.0E-08	1.000	2.3E-08	1.5E-08	1.0E-08	7.7E-09	7.0E-09

a Deposition 10%:20%:40% (bronchial:bronchiolar:alveolar-interstitial), 1.7 day retention half-time (ICRP Publication 68)

Table A.4. Effective dose rates for exposure of adults to inert gases

Nuclide	Physical half-life	Effective dose rate per unit air concentration (Sv d^{-1}/Bq m^{-3})
Argon		
Ar-37	35.0 d	4.1E-15
Ar-39	269 y	1.1E-11
Ar-41	1.83 h	5.3E-09
Krypton		
Kr-74	11.5 min	4.5E-09
Kr-76	14.8 h	1.6E-09
Kr-77	74.7 min	3.9E-09
Kr-79	1.46 d	9.7E-10
Kr-81	2.10E+05 y	2.1E-11
Kr-83m	1.83 h	2.1E-13
Kr-85	10.7 y	2.2E-11
Kr-85m	4.48 h	5.9E-10
Kr-87	1.27 h	3.4E-09
Kr-88	2.84 h	8.4E-09
Xenon		
Xe-120	40.0 min	1.5E-09
Xe-121	40.1 min	7.5E-09
Xe-122	20.1 h	1.9E-10
Xe-123	2.08 h	2.4E-09
Xe-125	17.0 h	9.3E-10
Xe-127	36.4 d	9.7E-10
Xe-129m	8.0 d	8.1E-11
Xe-131m	11.9 d	3.2E-11
Xe-133m	2.19 d	1.1E-10
Xe-133	5.24 d	1.2E-10
Xe-135m	15.3 min	1.6E-09
Xe-135	9.10 h	9.6E-10
Xe-138	14.2 min	4.7E-09

ANNEXE B. INHALATION DOSE COEFFICIENTS FOR WORKERS EXPOSED TO ^{226}Ra

The general approach in *ICRP Publication 68* was, with the exception of some isotopes of noble gases, that decay products produced in the respiratory tract have the same biokinetic behaviour as their parent nuclides. This also applied to the radon isotopes. This assumption appears to be overly cautious for ^{226}Ra, where a substantial fraction of the ^{222}Rn decay products would be expected to escape from the lung. Therefore, radon isotopes produced in the respiratory tract are now assumed to be removed to the environment with a removal rate of 100 d^{-1}. This new assumption influences significantly only the dose coefficients for ^{226}Ra. The revised values are 3.2×10^{-6} Sv Bq^{-1} and 2.2×10^{-6} Sv Bq^{-1} for an AMAD of 1 and 5 μm, respectively.